The Inverse Gaussian
Distribution

STATISTICS: Textbooks and Monographs

A Series Edited by

D. B. Owen, Coordinating Editor
Department of Statistics
Southern Methodist University
Dallas, Texas

R. G. Cornell, Associate Editor
for Biostatistics
University of Michigan

W. J. Kennedy, Associate Editor
for Statistical Computing
Iowa State University

A. M. Kshirsagar, Associate Editor
for Multivariate Analysis and
Experimental Design
University of Michigan

E. G. Schilling, Associate Editor
for Statistical Quality Control
Rochester Institute of Technology

ADDITIONAL VOLUMES IN PREPARATION

The Inverse Gaussian Distribution

Theory, Methodology, and Applications

Raj S. Chhikara

*University of Houston
  at Clear Lake
Houston, Texas*

J. Leroy Folks

*Oklahoma State University
Stillwater, Oklahoma*

 CRC Press
Taylor & Francis Group
Boca Raton London New York

CRC Press is an imprint of the
Taylor & Francis Group, an **informa** business

First published 1989 by Marcel Dekker, Inc.

Published 2019 by CRC Press
Taylor & Francis Group
6000 Broken Sound Parkway NW, Suite 300
Boca Raton, FL 33487-2742

© 1989 by Taylor & Francis Group, LLC
CRC Press is an imprint of Taylor & Francis Group, an Informa business

First issued in paperback 2019

No claim to original U.S. Government works

ISBN 13: 978-0-367-45126-4 (pbk)
ISBN 13: 978-0-8247-7997-9 (hbk)

Visit the Taylor & Francis Web site at
http://www.taylorandfrancis.com

and the CRC Press Web site at
http://www.crcpress.com

Library of Congress Cataloging-in-Publication Data

Chhikara, Raj S.
    The inverse Gaussian distribution.

    (Statistics, textbooks and monographs; v. 95)
    Bibliography: p.
    Includes index.
    1. Inverse Gaussian distribution. I. Folks,
J. Leroy. II. Title. III. Series.
QA276.7.C48 1988 519.5   88-20271
ISBN 0-8247-7997-5

Preface

This monograph has been prepared to consolidate research on the inverse Gaussian distribution. Although a large body of published work on the distribution exists, it still remains obscure to many statisticians and other users of statistical distributions. Because the inverse Gaussian is a natural choice for modeling statistical behavior in many applications, the extensive description of the distribution and related work presented here should prove to be of value.

We have emphasized the presentation of the statistical properties, methods, and applications of the two-parameter inverse Gaussian family of distributions. Because of the many useful sampling and statistical analysis results available for this family, we hope to see the inverse Gaussian distribution considered for further research and applications. The extensive bibliography given at the end covers what is to our knowledge the entire published work to date on the inverse Gaussian and its related topics.

We especially wish to express our gratitude to Professor G. P. Patil for his encouragement. His continued interest in this work served as motivation in our pressing on to its completion. At one time this volume was to be one of a series resulting from a NATO advanced Study Institute held at Trieste, Italy, July 10 to August 1, 1980. Professor Patil obtained extensive reviews on an earlier version of the manuscript. We are greatly indebted to Professor Patil and these reviewers.

Raj S. Chhikara
J. Leroy Folks

Contents

The Inverse Gaussian
Distribution

1

Introduction

1.1 BACKGROUND AND PURPOSE OF THE STUDY

The inverse Gaussian distribution and its possible usefulness in statistics first came to the attention of the authors about 15 years ago. Since that time they have been involved in studying the distribution and in developing statistical methods for data supposed to have arisen from the inverse Gaussian. Much progress has been made by a number of people, and several sizable works reviewing the inverse Gaussian exist. Wasan (1969a) reviewed the early work and discussed certain limiting forms and characterizations of the inverse Gaussian distribution. Johnson and Kotz (1970) summarized its statistical properties known up to that time. Folks and Chhikara (1978) reviewed the close analogy between the statistical properties of the inverse Gaussian and those of the normal distribution. They also discussed many useful statistical methods based on the inverse Gaussian distribution. In the past decade a significant number of papers has been published on the topic. Because of the sizable amount of work being done on this distribution, there is a need to bring together in the form of a monograph much of the inverse Gaussian research oriented toward statistical inference and methodology.

The purpose of this book is to summarize as completely as they are presently known the properties of the inverse Gaussian distribution and its uses in theoretical and applied statistics. The authors hope that such a summary will stimulate additional research, resulting in the solution of many interesting yet currently unsolved problems.

1.2 HISTORICAL REVIEW OF THE INVERSE GAUSSIAN DISTRIBUTION

In 1828 Robert Brown (1773–1858), one of England's greatest botanists, wrote a pamphlet describing observations made during three months in 1827. Brown was interested in studying the action of pollen in the impregnation process. Beginning with the pollen of *Clarkia puchella*, he found a swimming, dancing motion of pollen particles when these particles were immersed in water. He broadened his research to a wide array of plant pollen, finding a similar motion in every case. He then decided to try dead material. "Having found motion in the particles of the pollen of all the living plants which I had examined, I was led to inquire whether this property continued after the death of the plant, and for what length of time it was retained."

Brown then examined particles from several dead plants and subsequently extended his observations to fossilized plant remains. Always he saw the mysterious motion of particles. He was led to look at mineral specimens of all sorts: "But hence I inferred that these molecules were not limited to organic bodies, nor even to their products." He examined virtually everything he could imagine, from the soot of London to a fragment from the Sphinx.

He apparently believed he had discovered a new type of particle, common to all matter, organic and inorganic. Several researchers before him had observed and noted the motion of microscopic organic particles in fluids, but his work led to the realization that it was a physical phenomenon, not a biological one. Whether he himself initially recognized it as a physical phenomenon is open to debate. Brush (1968) seems to feel that Brown was misunderstood because of Brown's informal style. MacDonald (1962), on the other hand, indicates that Brown initially drew the conclusion that he had encountered some elementary form of life in all matters.

During the rest of the century many experiments were made and much theorizing was done about the nature of the so-called Brownian motion.

In 1900 Bachelier derived the normal distribution as the law governing the position of a single grain performing one-dimensional Brownian motion. His work was similar to the modern theory of Brownian motion, but he lacked the idea of a measure on the path space which was supplied by Wiener in 1923.

Einstein (1905) also derived the normal distribution as the model for Brownian motion and the theory of Brownian movement was firmly launched.

Schrödinger (1915) considered Brownian motion with a positive drift and obtained the distribution of the first passage time. Smoluchowski (1915) also obtained this distribution.

The next emergence of the first passage time distribution can be traced to Tweedie (1941) and Wald (1944). Tweedie, attempting to extend Schrödinger's results, was led to notice the inverse relationship between the cumulant-generating function of the time to cover unit distance and the cumulant-generating function of the distance covered in unit time. Tweedie (1945) also noted this type of relationship between the binomial and the negative binomial, and between the Poisson and the exponential distribution. He proposed to call them inverse statistical variates. Later, in 1956, he used the name inverse Gaussian for the first passage time distribution of the Brownian motion. He published a detailed study of the distribution in 1957 which established many of its important statistical properties. Folks and Chhikara (1978) suggested that because of the importance of Tweedie's work it might be more appropriate to call the distribution Tweedie's distribution.

A special case of the distribution was given by Wald in 1947. Wald derived it as an approximation of the sample size distribution in a sequential probability ratio test. The distribution is sometimes known as Wald's distribution, particularly in the Russian literature.

1.3   ANALOGIES WITH NORMAL DISTRIBUTION THEORY

A study of sampling distributions from the inverse Gaussian distribution has revealed many results which are analogous to the sampling results known for the normal distribution. To anticipate a little we mention the following: Consider a random sample X_1, $X_2, \ldots, X_n$ from an inverse Gaussian distribution. Then just as in the normal case,

1.   The sample mean $\bar{X}$ is inverse Gaussian.
2. $\bar{X}$ and $\Sigma(1/X_i - 1/\bar{X})$ are independently distributed statistics.
3.   The term in the exponent of the distribution is $(-\frac{1}{2})$ times a chi-square variable.
4.   The uniformly most powerful unbiased test for the mean employs Student's t distribution.

These results, including the others discussed in this work, are surprising, sometimes even startling, when first encountered. Although the authors have been able to develop many analogous results, they do not understand at this writing the underlying reasons. There is no general sampling rule or transformation by which the inverse Gaussian variable can be obtained from the normal.

1.4   APPLICATIONS OF THE INVERSE GAUSSIAN DISTRIBUTION

The notion of Brownian motion is applicable in describing the inherent process of many phenomena, particularly in the natural and physical sciences. Because the first passage time of a Brownian motion is distributed as inverse Gaussian, it is logical to use it as a lifetime model. We give several examples of lifetime in this mono-graph and discuss certain useful properties of the inverse Gaussian model in studying the life testing and reliability of a product or device. For example, the failure function for the inverse Gaussian is shown to be nonmonotonic, where it first increases and then decreases, approaching a constant value as the lifetime becomes infinite. This property is often found in cases of lifetimes dominated by early occurrence of an event. An example of repair times for an airborne communication transceiver given in Chapter 9 illustrates this property, and thus the usefulness of the inverse Gaussian as a repair time model.

Besides the field of reliability, the inverse Gaussian distribution has been used in a wide range of applications. Again, most of these applications are based on the idea of first passage time for an underlying process. Discussed in Chapter 10 are examples from many diverse fields such as cardiology, hydrology, demography, linguistics, employment service, labor disputes, and finance.

One of many interesting properties of the inverse Gaussian is that the family of inverse Gaussian distributions is fairly wide. As can be easily seen from the sketches of its probability density function given in Chapter 2, the inverse Gaussian distribution can represent a highly skewed to an almost normal distribution. We have given an example from linguistics showing the applicability of the inverse Gaussian in studying word frequency distribution, which is expected to be highly skewed. Other applications involving skewed distributions of wind energy and agricultural fields are also discussed in Chapter 10.

1.5 PROBLEMS OF DATA ANALYSIS WITH SKEWED DISTRIBUTIONS

In many areas of statistical application, handling of skewed data is by no means an exception but a fact of life. However, standard statistical methods for the normal distribution are commonly used for the data analysis. This is primarily due to lack of alternative methods that are easily available and also easy to understand.

Although the lognormal, gamma, and Weibull distributions enjoy extensive use in certain special areas, none of them allow for a wide range of statistical methods comparable to those based on the normal distribution, such as one- and two-sample t tests, analysis of variance, confidence intervals, regression analysis, and so on. Although some exact sampling results exist for all of these distributions, they have not lent themselves to the development of a comprehensive statistical methodology for skewed data analyses. For this reason investigators most often assume the normal distribution to analyze their data unless the choice of a different distribution is made obvious by the physical characteristics. That is perhaps why the gamma and Weibull distributions are extensively used in reliability studies but very little in experimental work, even though they may be appropriate for applications based, for example, on goodness of fit criteria.

When confronted with skewed distributions, investigators usually resort to a transformation in order to normalize the data. For example, the Box & Cox transformation (1964) is put forward partially because of the desire to eliminate skewness in data. Although it may be true, for example, that the reciprocal of the response variable is better described by an experimental design model than the response variable itself, there is still a major problem of interpretation involved when we consider the data analysis using the transformed variable. If possible, it is desirable to analyze the data as observed using statistical methods based on skewed distributions. The authors feel that the application of the inverse Gaussian when appropriate can meet part of this need for skewed data analysis.

2

Properties of the Inverse Gaussian Distribution

2.1 INTRODUCTION

In this chapter we give the probability density function (pdf) of the inverse Gaussian distribution and discuss its statistical properties. We primarily consider the pdf form due to Tweedie (1957a) because this leads to the development of a close analogy between the statistical properties of the inverse Gaussian and those of the normal distribution. The discussion and results given in the next two sections were obtained by Tweedie (1957a), who laid the foundation for research work on the inverse Gaussian distribution.

2.2 PROBABILITY DENSITY FUNCTION

The pdf of an inverse Gaussian distribution random variable X is

$$f(x; \mu, \lambda) = \sqrt{\frac{\lambda}{2\pi}} \, x^{-3/2} \, \exp\left(-\frac{\lambda(x - \mu)^2}{2\mu^2 x}\right), \qquad x > 0 \qquad (2.1)$$

where $\mu > 0$ and $\lambda > 0$. The parameter μ is the mean of the distribution and λ is a scale parameter. Tweedie listed three other forms of (2.1), which he obtained by replacing the set of parameters (μ, λ) by (α, λ), or (μ, ϕ), or (ϕ, λ) using the relationship given by $\mu = \lambda/\phi = (2\alpha)^{-1/2}$. Each of these forms was found useful by him in his investigation of Brownian motion for the colloid particles in a Tuorila electrophoretic cell. Both μ and λ are of the same physical dimensions as the random variable X itself; but the parameter $\phi = \lambda/\mu$ is invariant under a scale transformation of X as can be seen from the following relationships:

$$f(x; \mu, \lambda) = \mu^{-1} f\left(\frac{x}{\mu}; 1, \phi\right) = \lambda^{-1} f\left(\frac{x}{\lambda}; \phi, 1\right). \tag{2.2}$$

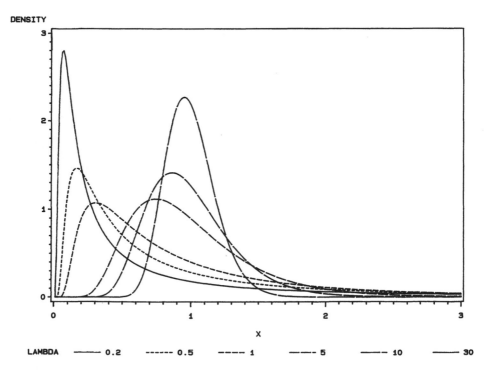

Figure 2.1   Inverse Gaussian densities with $\mu = 1$ for six values of λ.

The probability density can be numerically computed using any of the three forms in (2.2). As given later in Section 2.5, the cumulative distribution function depends essentially on only two variables, which might be taken as x/μ and ϕ. Accordingly, the case $\mu = 1$ for the (μ, ϕ) parametric form of (2.2) could be adopted as a standard form. This has also been obtained as a limiting form of the distribution of the sample size in a Wald's sequential probability ratio test and is sometimes referred to as the *standard Wald's distribution* (Zigangirov, 1962).

The shape of the distribution depends on ϕ only; hence ϕ is the shape parameter. The inverse Gaussian density function represents a wide class of distributions, ranging from a highly skewed distribution to a symmetrical one as ϕ varies from 0 to ∞. The density curves shown in Figures 2.1 and 2.2 illustrate this property.

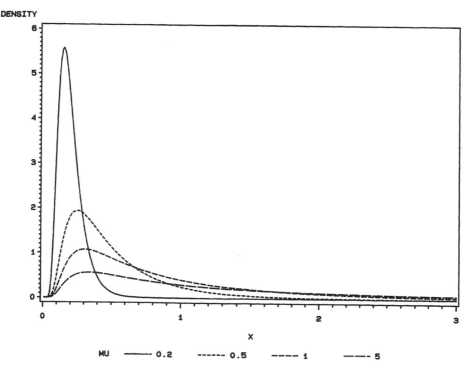

Figure 2.2 Inverse Gaussian densities with $\lambda = 1$ for four values of μ.

These curves are obtained by specifying $\mu = 1$ and varying λ or ϕ in Figure 2.1, and by specifying $\lambda = 1$ and varying μ or ϕ in Figure 2.2. The density function is unimodal, with its mode at

$$\mu\left[\left(1 + \frac{9}{4\phi^2}\right)^{1/2} - \frac{3}{2\phi}\right].$$

Henceforth in this work the expression (2.1) will be regarded as a standard form of the inverse Gaussian distribution with parameters μ and λ, and the random variable X distributed according to formula (2.1) will be denoted by $X \sim \mathrm{IG}(\mu, \lambda)$. The abbreviation IG for the inverse Gaussian will be used in the monograph.

The family of inverse Gaussian distributions constitutes a natural exponential family of order 2. Letting $\psi = \lambda/\mu^2$, the pdf in (2.1) can be written as

$$f(x; \lambda, \psi) = \left(\frac{\lambda}{2\pi}\right)^{1/2} e^{\lambda\psi/2} x^{-3/2} e^{-(\lambda x^{-1} + \psi x/2)}, \qquad (2.3)$$

which is of the natural exponential model form

$$a(\theta)b(x)e^{\theta \cdot t(x)},$$

where $\theta = (\lambda, \psi)$, $t(x) = (-\frac{1}{2})(x^{-1}, x)$, $b(x) = x^{-3/2}$, and $a(\theta) = (\lambda/2\pi)^{1/2} \exp((\lambda\psi)^{1/2})$. The canonical parameter domain is

$$\otimes = \left\{\theta \in R^2, \int x^{-3/2} \exp(-\tfrac{1}{2}\theta t(x))\, dx < \infty\right\}$$

$$= (0, \infty) \times [0, \infty),$$

showing that the distributions with $\psi = 0$ also belong to the family (Barndorff-Nielsen, 1978b). Note that if the parameters (λ, μ) are used as in (2.1), the family does not include the distributions with $\mu = \infty$ (i.e., $\psi = 0$). That these distributions naturally belong to the family is evident since they are the first passage time distributions for a Brownian motion with zero drift, as discussed in Chapter 3. Also, these are the *stable distributions* with characteristic exponent or index $\frac{1}{2}$ (Feller, 1966).

2.3 CHARACTERISTIC FUNCTION AND MOMENTS

The characteristic function of $X \sim \text{IG}(\mu, \lambda)$,

$$E[e^{itX}] = \int_0^\infty e^{itx} f(x; \mu, \lambda)\, dx$$

can be easily obtained by combining the two integrand terms and simplifying the right side in the form

$$E[e^{itX}] = \exp\left\{ \frac{\lambda}{\mu}\left[1 - \left(1 - \frac{2it\mu^2}{\lambda}\right)^{1/2}\right]\right\}$$
$$\times \int_0^\infty f\left(x; \mu\left(1 - \frac{2it\mu^2}{\lambda}\right)^{-1/2}, \lambda\right) dx.$$

Thus the characteristic function, denoted by $C_X(t)$, is given by

$$C_X(t) = \exp\left\{ \frac{\lambda}{\mu}\left[1 - \left(1 - \frac{2i\mu^2 t}{\lambda}\right)^{1/2}\right]\right\}. \tag{2.4}$$

All positive and negative moments exist. The positive moments can be obtained by differentiating the characteristic function in (2.4) and the negative moments by integrating it (Cressie et al., 1981). Taking the rth derivative of $C_X(t)$ and evaluating it at $t = 0$, we get

$$E[X^r] = \mu^r \sum_{s=0}^{r-1} \frac{(r-1+s)!}{s!(r-1-s)!} \left(2\frac{\lambda}{\mu}\right)^{-s}. \tag{2.5}$$

The first four moments about zero are

$$\mu$$

$$\mu^2 + \frac{\mu^3}{\lambda}$$

$$\mu^3 + 3\frac{\mu^4}{\lambda} + 3\frac{\mu^5}{\lambda^2}$$

$$\mu^4 + 6\frac{\mu^5}{\lambda} + 15\frac{\mu^6}{\lambda^2} + 15\frac{\mu^7}{\lambda^3}. \tag{2.6}$$

The central moments can be obtained either from (2.5) and (2.6) or from the power series expansion of the cumulant-generating function (cgf) given by

$$\ln E(e^{tX}) = \frac{\lambda}{\mu}\left[1 - \left(1 - \frac{2\mu^2 t}{\lambda}\right)^{1/2}\right], \qquad t < \frac{\lambda}{2\mu^2}. \tag{2.7}$$

For the central moments it is convenient to work with cumulants K_r, where K_r is the coefficient of $(t^r/r!)$ in the expansion of a cgf, since these (except for K_1) are invariant for translations of random variables. It easily follows from (2.7) that

$$K_1 = \mu$$

and

$$K_r = 1 \cdot 3 \cdot 5 \cdots (2r - 3)\mu^{2r-1}\lambda^{1-r}, \qquad r \geq 2. \tag{2.8}$$

From (2.6) or (2.8), the second, third, and fourth central moments are

$$\mu_2 = \frac{\mu^3}{\lambda}$$

$$\mu_3 = 3\frac{\mu^5}{\lambda^2}$$

$$\mu_4 = 15\frac{\mu^7}{\lambda^3} + 3\frac{\mu^6}{\lambda^2}. \tag{2.9}$$

The coefficient of variation of X is $\sqrt{\mu/\lambda}$. The measures of skewness and kurtosis are

$$\sqrt{\beta_1} = 3\sqrt{\frac{\mu}{\lambda}}$$

$$\beta_2 = 15\left(\frac{\mu}{\lambda}\right) + 3, \tag{2.10}$$

respectively. Since $\mu/\lambda = \phi^{-1}$, the IG distribution becomes more and more nearly normal when ϕ is increased.

The negative moments are given later in Section 4.2, which contains the distribution of X^{-1}. We may, however, point out that there is a remarkably simple relationship between positive and negative moments given by

$$E[X^{-r}] = \frac{E[X^{r+1}]}{\mu^{2r+1}} . \tag{2.11}$$

2.4 SOME USEFUL PROPERTIES

The inverse Gaussian distribution enjoys certain reproductive and other useful properties. A few of these properties are discussed next.

1.   For any number $c > 0$, the characteristic function of cX is $\exp\{(\lambda/\mu)[1 - (1 - 2it\mu^2 c/\lambda)^{1/2}]\}$. Thus, cX is inverse Gaussian distributed with parameters $c\mu$ and $c\lambda$. So the family of inverse Gaussian distributions is closed under a change of scale. Unfortunately, the property of reproducibility does not hold with respect to a change of location.

2.   For a linear combination $\Sigma c_i X_i$, $c_i > 0$, where $X_i \sim IG(\mu_i, \lambda_i)$, it follows easily that if $\lambda_i/(\mu_i^2 c_i) = \xi$ for all i, then the characteristic function of $\Sigma c_i X_i$ is

$$\exp\left\{\xi(\Sigma c_i\mu_i)\left[1 - \left(1 - 2\frac{it}{\xi}\right)^{1/2}\right]\right\}.$$

Thus $\Sigma c_i X_i \sim IG(\Sigma c_i\mu_i, \xi(\Sigma c_i\mu_i)^2)$. The constancy of $\lambda_i/(\mu_i^2 c_i)$ is a necessary condition for $\Sigma c_i X_i$ to be distributed as IG with parameters $\Sigma c_i\mu_i$ and $\xi(\Sigma c_i\mu_i)^2$. This is easily established by equating the cgf of $\Sigma c_i X_i$ and the sum of cgf's of $c_i X_i$, $i = 1, 2, \ldots, n$. So the additive property of the inverse Gaussian is restricted by a required relationship between the two parameters.

3.   Wasan (1968) showed that the family of inverse Gaussian distributions is *complete*; namely,

$$\int_0^\infty h(x)f(x; \mu, \lambda)\, dx = 0 \tag{2.12}$$

implies that $h(x) = 0$ almost everywhere. Simplifying the integrand in (2.12), it follows that

$$\int_0^\infty h(x)f(x;\mu,\lambda)\,dx = \int_0^\infty \left[h(x)\left(\frac{\lambda}{2\pi x^3}\right)^{1/2} e^{\lambda/\mu - \lambda/2x} \right] e^{-vx}\,dx,$$

$$(2.13)$$

where $v = \lambda/2\mu^2$, which is the Laplace transform of the function given in the bracket. Denote

$$m(x) = \left(\frac{\lambda}{2\pi x^3}\right)^{1/2} e^{\lambda/\mu - \lambda/2x}.$$

Hence it follows from (2.12) and (2.13) that $h(x)m(x) = 0$ almost everywhere. Since $m(x) > 0$ for $x > 0$, one has $h(x) = 0$ almost everywhere. The completeness property is useful in developing inferential properties for the distribution.

  4. The density function in (2.1) can be written as

$$f(x;\theta_1,\theta_2) = \sqrt{\frac{\theta_1}{2\pi}} \exp(\sqrt{\theta_1\theta_2})x^{-(3/2)} \exp[-\tfrac{1}{2}(\theta_1 x^{-1} + \theta_2 x)],$$

which represents an exponential family of distributions with parametric vector $\theta = -\tfrac{1}{2}(\theta_1, \theta_2)$ and statistic $T = (X^{-1}, X)$. Moreover, this is *reproductive* in variable X because the mean $\bar{X}$ of a sample $X_1, X_2, \ldots, X_n$ from (2.1) has the inverse Gaussian distribution. See Barndorff-Nielsen and Blaesild (1983b) for a definition and many other useful properties of the reproductive exponential family.

2.5 DISTRIBUTION FUNCTION

For $X \sim IG(\mu, \lambda)$, Seshadri (unpublished) proved by finding the moment-generating function that $\lambda(X - \mu)^2/\mu^2 X$ is distributed as chi square with one degree of freedom. It should be noted from the density function in (2.1) that the new variable is the negative of twice the term in the exponent; hence an analogy with the normal is established, as mentioned previously in Chapter 1.

Based on this transformation Shuster (1968) expressed the cumulative distribution function $F(x)$ of X in terms of the standard normal distribution function, Φ, given by

$$F(x) = \Phi\left[\sqrt{\frac{\lambda}{x}}\left(\frac{x}{\mu} - 1\right)\right] + e^{2\lambda/\mu}\Phi\left[-\sqrt{\frac{\lambda}{x}}\left(1 + \frac{x}{\mu}\right)\right].$$

(2.14)

His proof is fairly complex and tedious. First he expressed it in terms of a chi-square cdf and then obtained (2.14). An independent and simple derivation of (2.14) for the special case $\mu = 1$ was given by Zigangirov (1962). Another simpler and direct derivation was given by Chhikara and Folks (1974) based on their following result.

Theorem 2.1   Let $Y = \sqrt{\lambda}(X - \mu)/\mu\sqrt{X}$. Then the pdf of Y is given by

$$g(y) = \left(1 - \frac{y}{\sqrt{4\lambda/\mu + y^2}}\right)\left(\frac{1}{\sqrt{2\pi}}e^{-y^2/2}\right), \qquad -\infty < y < \infty.$$

(2.15)

Proof: The transformation $y = \sqrt{\lambda}(x - \mu)/\mu\sqrt{x}$ is one-to-one and as x varies from 0 to ∞, y varies from $-\infty$ to ∞. The result in (2.15) follows after inverting the transformation and then making appropriate substitutions for $f(x; \mu, \lambda)\, dx/dy$.

It can be shown [see Chhikara and Folks (1974) for details] that the cumulative distribution function $G(y)$ of Y is

$$G(y) = \Phi(y) + e^{2\lambda/\mu}\Phi\left(-\sqrt{y^2 + \frac{4\lambda}{\mu}}\right), \qquad -\infty < y < \infty.$$

(2.16)

Since y is a one-to-one transformation, $F(x) = G(y)$; hence the result in (2.14) follows.

It follows from (2.16) that $G(y) \to \Phi(y)$ as $\phi = \lambda/\mu \to \infty$. Because of this and because of the one-to-one relationship between x and y, one finds that the distribution of X is asymptotically normal with mean μ and variance μ^3/λ. This can also be seen from (2.10) where the skewness goes to zero and the kurtosis to 3 as $\phi \to \infty$. Asymptotic normality was first given by Wald (1947).

It may be pointed out that the transformation in Theorem 2.1 establishes a basic relationship between IG and the normal. As will be seen later, it helps in making inferences for the inverse Gaussian distribution by using existing tables and results for the normal.

2.6   STANDARDIZATION OF INVERSE GAUSSIAN

No transformation exists for an inverse Gaussian random variable that leads to a standard form for the distribution which is similar to the standard normal distribution. However, if a change of scale is considered, it is possible to reduce the number of parameters and to obtain an inverse Gaussian distribution with a single parameter. We state the result in the following theorem.

Theorem 2.2   Let $Z = \lambda X/\mu^2$. Then $Z \sim IG(\phi, \phi^2)$ where $\phi = \lambda/\mu$. The proof is straightforward.

Wasan and Roy (1969) have tabulated the percentage points of the distribution of Z for different values of ϕ. Unfortunately, these tables are not accurate, as pointed out by Whitmore and Yalousky (1978).

The transformed variable Y in Theorem 2.1 can be written in terms of Z as

$$Y = \frac{Z - \phi}{\sqrt{Z}}. \tag{2.17}$$

The density function of Y given in (2.15) depends on the parameter ϕ. It is monotonically decreasing for $y < 0$ and increasing for $y > 0$

as ϕ increases; further, as noted earlier, it approaches the standard normal density as $\phi \to \infty$. Its curves for some values of ϕ are given in Figure 2.3.

Although the transformation Y provides us with a distribution that resembles the standard normal, it still cannot be regarded as a normalized form in terms of zero mean and unit variance unless ϕ is large.

Another normalization of the inverse Gaussian random variable is given by Whitmore and Yalovsky (1978) using the following logarithmic transformation:

$$W = \frac{1}{2\sqrt{\phi}} + \sqrt{\phi} \, \log \frac{X}{\mu}. \qquad (2.18)$$

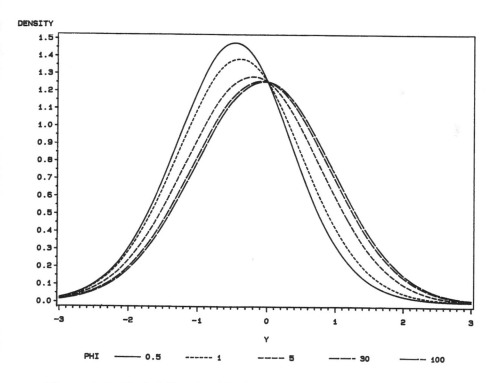

Figure 2.3 Probability densities for Y for five values of ϕ.

The density function of W is

$$h(\omega) = \frac{1}{\sqrt{2\pi}} e^{-\omega^2/2} \exp\left[\frac{1}{8\phi} - \frac{\phi}{2} \sum_{k=2}^{\infty} \left(\frac{\omega}{\sqrt{\phi}} - \frac{1}{2\phi}\right)^{2k} \frac{1}{(2k)!}\right].$$

(2.19)

When ϕ is reasonably large, the second part on the right side in (2.19) is approximately one, and thus W is distributed approximately as the standard normal.

Since the first term on the right side of equation (2.18) becomes negligible as ϕ becomes large, one can alternatively consider the random variable

$$\sqrt{\phi} \log \frac{X}{\mu}$$

(2.20)

for a normalization of X.

Whitmore and Yalovsky (1978) also pointed out that the variable

$$\sqrt{\phi}\left(\frac{X}{\mu} - 1\right)$$

(2.21)

is asymptotically normal as $\phi \to \infty$, and showed that the distribution of variable W defined in (2.18) tends faster to normality than the distribution of $\sqrt{\phi}(X/\mu - 1)$. Of course, the same result holds true for the variable in (2.20). However, no such comparison has been made with respect to the variable Y defined in (2.17).

2.7   A THREE-PARAMETER INVERSE GAUSSIAN DISTRIBUTION

The lack of the invariance property under change of location makes it desirable to introduce an additional threshold parameter for the family of inverse Gaussian distributions. If there is a threshold value θ, a three-parameter inverse Gaussian distribution can be defined by assuming that $(X - \theta) \sim IG(\mu, \lambda)$, and that the density function of X is given by

$$f(x; \theta, \mu, \lambda) = \left[\frac{\lambda}{2\pi(x - \theta)^3}\right]^{1/2} \exp\left(-\frac{\lambda[(x - \theta) - \mu]^2}{2\mu^2(x - \theta)}\right),$$

(2.22)

$$-\infty < \theta < \infty, \mu > 0, \lambda > 0, x > \theta$$

For (2.22), $E(X) = \theta + \mu$, but $\lambda/\mu = \phi$ is a shape parameter as in the case of two-parameter inverse Gaussian distribution. The distribution given by the pdf in (2.22) provides an alternative to the three-parameter lognormal, gamma, Weibull, and other skewed distributions.

Chan et al. (1983) constructed tables of the pdf and cdf for the standardized form of (2.22) with zero mean and unit variance. Letting

$$Z = \frac{X - E(X)}{[\text{Var}(X)]^{1/2}},$$

the density function of Z is given by

$$g(z; \theta) = [2\pi(1 + \phi^{-1/2}z)^3]^{-1/2} \exp\left[\frac{-z^2}{2(1 + \phi^{-1/2}z)}\right],$$

(2.23)

where $z > -\phi^{1/2}$, and zero elsewhere. The corresponding cdf obtained by integrating (2.23) simplifies as

$$G(z; \phi) = \Phi\left(\frac{z}{(1 + \phi^{-1/2}z)^{1/2}}\right) + e^{2\phi}\Phi\left(\frac{-[z + 2\phi^{1/2}]}{(1 + \phi^{-1/2}z)^{1/2}}\right)$$

(2.24)

where Φ denotes the cdf for the standard normal. The two expressions in (2.23) and (2.24) are functions of $\phi = \lambda/\mu$ alone and can be easily computed by specifying ϕ. The tables in Chan et al. (1983) are given in terms of the third standardized moment (skewness), which they denote by α_3. Since $\alpha_3 = 3\phi^{-1/2}$, their tables can be directly used to obtain values of $g(z; \phi)$ given in (2.23) and $G(z; \phi)$ as in (2.24).

2.8 A GENERALIZED INVERSE GAUSSIAN DISTRIBUTION

We introduce a general class of distributions of which the inverse Gaussian is a special case. A three-parameter family of such distributions is given by the pdf

$$f(x; \gamma, \mu, \lambda) = \left[2\mu^\gamma K_\gamma\left(\frac{\lambda}{\mu}\right)\right]^{-1} x^{\gamma-1} \exp\left(-\frac{[\lambda x^{-1} + (\lambda/\mu^2)x]}{2}\right),$$
$$-\infty < \gamma < \infty, \mu > 0, \lambda \geqslant 0, x > 0 \qquad (2.25)$$

where $K_\gamma(\cdot)$ is the modified Bessel function of the third kind of order γ. This represents a wide class of distributions, including some well-known ones. When $\gamma = -(1/2)$, we obtain the inverse Gaussian distribution with pdf given in (2.1). This easily follows because

$$K_{\pm 1/2}\left(\frac{\lambda}{\mu}\right) = \left(\frac{\pi\mu}{2\lambda}\right)^{1/2} e^{-\lambda/\mu}. \qquad (2.26)$$

One may refer to Watson (1966) for this and other related expressions for the Bessel functions.

Jorgensen (1980) and others have called the distribution given by (2.25) the generalized inverse Gaussian. This name is used perhaps because the pdf in (2.25) is of the form of the inverse Gaussian pdf in (2.1), but it has an additional parameter which generalizes the latter.

Other special cases of (2.25) are the gamma distribution ($\lambda = 0$, $\gamma > 0$), the inverted gamma distribution ($\lambda/\mu^2 = 0$, $\gamma < 0$), the distribution of a reciprocal inverse Gaussian variate ($\gamma = \frac{1}{2}$), the hyperbola distribution ($\gamma = 0$) and the hyperbolic distribution ($\gamma = 1$). See Barndorff-Nielsen (1987b) for a description of the hyperbola and hyperbolic distributions. Also, one may refer to Barndorff-Nielsen and Blaesild (1980).

The characteristic function of X is

$$C_X(t) = \frac{K_\gamma((\lambda/\mu)(1 - 2\mu^2 it/\lambda)^{1/2})}{K_\gamma(\lambda/\mu)(1 - 2\mu^2 it/\lambda)^{\gamma/2}} \tag{2.27}$$

and the moments $\mu'_k = E(X^k)$ are given by

$$\mu'_k = \frac{K_{\gamma+k}(\lambda/\mu)}{K_\gamma(\lambda/\mu)} \mu^k. \tag{2.28}$$

The density function in (2.25) is unimodal. Some properties of the generalized inverse Gaussian seem to parallel those of the inverse Gaussian: If the random variable X is distributed according to (2.25), denoted by $X \sim \text{GIG}(\gamma, \mu, \lambda)$, then $X^{-1} \sim \text{GIG}(-\gamma, \lambda/\mu^2, \lambda)$, and for $c > 0$, $cX \sim \text{GIG}(\gamma, c\mu, c\lambda)$. The distributions of $(\lambda/\mu)^{1/2}(X/\mu - 1)$ and $(\lambda/\mu)^{1/2} \log X/\mu$ are asymptotically standard normal as $\lambda/\mu \to \infty$. The convergence is shown to be faster in the latter case (Jorgensen, 1980).

Besides the fact that the several known families of distributions belong to the generalized inverse Gaussian family, it is also useful as a mixing distribution in situations where there is a need for skewed weighting of distributions. For example, in the weighting of occurrence distributions for species, where a few species may be in great abundance and some very rare, a heavy tailed distribution as in (2.25) can be appropriately used for a weighting distribution. Good (1953) suggested its use for the mixing distribution with Poisson in connection with species sampling; but he did not use it because of its analytical intractability. Sichel (1974, 1975) found it very useful in constructing compound Poisson distributions in his investigations of sentence length and word frequency distributions. For details on the generalized inverse Gaussian distribution, refer to Jorgensen (1980), who has discussed its many inferential properties and applications.

A multivariate version of the generalized inverse Gaussian distribution is introduced in Barndorff-Nielsen et al. (1982). It includes the Wishart distribution as well as the inverse of a Wishart distribution as special cases.

3

Genesis

3.1 INTRODUCTION

As mentioned previously in Chapter 1, the inverse Gaussian distribution has its origin in the Brownian motion as a first passage time distribution. We first obtain it considering a Wiener process with both positive and negative drift. The case of zero drift is also discussed. Wald's distribution for the sample size in a sequential probability ratio test is then described. Following Tweedie (1945) we then consider the random variables which are inversely related in terms of their cumulant-generating functions and their means.

3.2 FIRST PASSAGE TIME IN A WIENER PROCESS

3.2.1 General Discussion

Let us define a Wiener process with drift v and variance σ^2 as a process $X(t)$ with the following properties:

1. $X(t)$ has independent increments; for any $t_1 < t_2 < t_3 < t_4$, $X(t_2) - X(t_1)$ and $X(t_4) - X(t_3)$ are independent.
2. $X(t_2) - X(t_1)$ is normally distributed with mean $v(t_2 - t_1)$ and variance $\sigma^2(t_2 - t_1)$ with $t_1 < t_2$.

Thus the probability density function for $X(t)$ given that the process started at x_0 is

$$f(x; x_0, t) = \frac{1}{\sigma\sqrt{2\pi t}} \exp\left(-\frac{(x - x_0 - vt)^2}{2\sigma^2 t}\right). \tag{3.1}$$

Let us now consider the first passage time T of $X(t)$ to $a > x_0$. In order for T to be the first passage time, we require

$$X(0) = x_0, \qquad X(t) < a, \, 0 < t < T$$

and

$$X(T) = a.$$

Following Prabhu (1965) we shall obtain the probability density function of T given that $T < \infty$ by finding its Laplace transform. Denote the conditional probability density function of T given that $T < \infty$ by $g(t; x_0, a)$. That is, let $f_{T|T<\infty} = g(t; x_0, a)$. Denote the Laplace transforms of f and g by

$$f^*(x; x_0, \theta) = \int_0^\infty e^{-\theta t} f(x; x_0, t)\, dt \tag{3.2}$$

$$g^*(\theta; x_0, a) = \int_0^\infty e^{-\theta t} g(t; x_0, a)\, dt. \tag{3.3}$$

Note that g^* is just a moment-generating function. That is, $g^*(-\theta; x_0, a) = E[e^{-\theta T} \mid T < \infty]$. On the other hand, f^* is not a moment-generating function of X, the integration being taken with respect to t.

The following theorem is used to find g^*.

Theorem 3.1   If $x_0 < a < x$, then

$$P(T < \infty)g^*(\theta; x_0, a) = \frac{f^*(x; x_0, \theta)}{f^*(x; a, \theta)}. \tag{3.4}$$

Proof: Refer to Figure 3.1. $X(t)$ can proceed to x from x_0 in time t, $x_0 < a < x$, by reaching a for the first time in time s and then reaching x from a in time $t - s$. Thus

$$f(x; x_0, t) = P(T < \infty) \int_0^t g(s; x_0, a) f(x; a, t - s)\, ds. \qquad (3.5)$$

Taking Laplace transforms of both sides of equation (3.5), we obtain

$$f^*(x; x_0, \theta) = P(T < \infty) \int_0^\infty g(s; x_0, a)$$

$$\times \int_s^\infty e^{-\theta t} f(x; a, t - s)\, dt\, ds. \qquad (3.6)$$

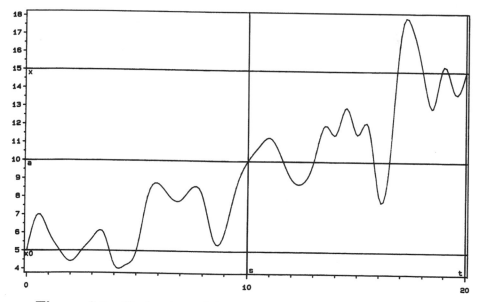

Figure 3.1 Derivation of first passage time distribution.

Let $\tau = t - s$. Then

$$
f^*(x; x_0, \theta) = P(T < \infty) \int_0^\infty e^{-\theta s} g(s; x_0, a)\, ds
$$

$$
\times \int_0^\infty e^{-\theta \tau} f(x; a, \tau)\, d\tau. \tag{3.7}
$$

So

$$
f^*(x; x_0, \theta) = P(T < \infty) g^*(\theta; x_0, a) f^*(x; a, \theta).
$$

Therefore equation (3.4) holds.

To apply the theorem, we must find the Laplace transform (3.2) of the probability density function (3.1) at x_0 and at a. Consider the negative of the term in the exponent of the integrand for $f^*(x; x_0, \theta)$, which is

$$
\theta t + \frac{(x - x_0)^2}{2\sigma^2 t} - \frac{v(x - x_0)}{\sigma^2} + \frac{v^2 t}{2\sigma^2}
$$

$$
= \frac{c}{2d^2 t}(t - d)^2 + \frac{x - x_0}{\sigma^2}\left(\sqrt{v^2 + 2\sigma^2 \theta} - v\right) \tag{3.8}
$$

where

$$
c = \frac{(x - x_0)^2}{\sigma^2} \qquad \text{and} \qquad d = \frac{x - x_0}{\sqrt{v^2 + 2\sigma^2 \theta}}. \tag{3.9}
$$

Then we can see that the Laplace transform f^* is given by

$$
f^*(x; x_0, \theta)
$$

$$
= \frac{1}{\sigma\sqrt{\lambda}}\left\{\exp\left[-\frac{x - x_0}{\sigma^2}\left(\sqrt{v^2 + 2\sigma^2\theta} - v\right)\right]\right\}
$$

$$
\times \int_0^\infty t\, \frac{c}{2\pi t^3} \exp\left[-\frac{c(t - d)^2}{2d^2 t}\right] dt
$$

$$
= \frac{d}{\sigma\sqrt{c}} \exp\left[\frac{x_0 - x}{\sigma^2}\left(\sqrt{v^2 + 2\sigma^2\theta} - v\right)\right] \tag{3.10}
$$

Then, using Theorem 3.1, stated earlier, we obtain

$$P(T < \infty)g^*(\theta; x_0, a) = \exp\left[\frac{x_0 - a}{\sigma^2}\left(\sqrt{v^2 + 2\sigma^2\theta} - v\right)\right].$$

(3.11)

3.2.2 Positive Drift

With $v > 0$, we obtain from (3.11) that

$$P(T < \infty)g^*(\theta; x_0, a) = \exp\left\{\frac{(a - x_0)v}{\sigma^2}\left[1 - \left(1 + 2\frac{\sigma^2\theta}{v^2}\right)^{1/2}\right]\right\}$$

$$= \exp\left\{\frac{\lambda}{\mu}\left[1 - \left(1 + \frac{2\mu^2\theta}{\lambda}\right)^{1/2}\right]\right\},$$

(3.12)

where

$$\mu = \frac{a - x_0}{v} \quad \text{and} \quad \lambda = \frac{(a - x_0)^2}{\sigma^2}.$$

When $\theta = 0$, the Laplace transform g^* equals 1 and therefore letting $\theta = 0$ in (3.12) gives us $P(T < \infty) = 1$. Also, because the moment-generating function $m_T(\theta) = E[e^{T\theta}] = g^*(-\theta; x_0, a)$, we have

$$m_T(\theta) = \exp\left\{\frac{\lambda}{\mu}\left[1 - \left(1 - \frac{2\mu^2\theta}{\lambda}\right)^{1/2}\right]\right\}.$$

(3.13)

This is, of course, just the moment-generating function of an inverse Gaussian distribution with parameters μ and λ related to the drift and diffusion parameters of Brownian motion by (3.12). Thus the first passage time T has the inverse Gaussian distribution with parameters as given by (2.1).

The derivation just given is only one of several ways of obtaining the first passage time distribution. Derivations more commonly given use either the backward or forward equations of diffusion processes (see Cox and Miller, 1965). However, given what we presently know about the inverse Gaussian, the presentation here seems more straightforward and suffices for depicting it to be the first passage time distribution in a Wiener process with positive drift.

3.2.3 Negative Drift

With $v < 0$, we obtain from (3.11) that

$$P(T < \infty)g^*(\theta; x_0, a) = \exp\left[\frac{(a - x_0)v}{\sigma^2}\left(1 + \sqrt{1 + \frac{2\sigma^2\theta}{v^2}}\right)\right].$$

$$(3.14)$$

From this we have

$$P(T < \infty)g^*(\theta; x_0, a) = \exp\left\{\frac{\lambda}{\mu}\left[1 + \left(1 + \frac{2\mu^2\theta}{\lambda}\right)^{1/2}\right]\right\}. \quad (3.15)$$

Letting $\theta = 0$ gives $P(T < \infty) = e^{2\lambda/\mu}$ and

$$g^*(\theta; x_0, a) = \exp\left\{-\frac{\lambda}{\mu}\left[1 - \left(1 + \frac{2\mu^2\theta}{\lambda}\right)^{1/2}\right]\right\}. \quad (3.16)$$

The conditional moment-generating function of T given that $T < \infty$, $E[E^{T\theta} | T < \infty] = g^*(-\theta; x_0, a)$, is

$$m_{T|T<\infty}(\theta) = \exp\left\{-\frac{\lambda}{\mu}\left[1 - \left(1 - \frac{2\mu^2\theta}{\lambda}\right)^{1/2}\right]\right\}. \quad (3.17)$$

We recognize this as the moment-generating function of an inverse Gaussian distribution $IG(-\mu, \lambda)$.

3.2.4 Zero Drift

With $v = 0$ we obtain from (3.11) that

$$P(T < \infty)g^*(\theta; x_0, a) = \exp\left(\frac{x_0 - a}{\sigma^2}\sqrt{2\sigma^2\theta}\right)$$

$$= \exp\left(-\sqrt{2\lambda\theta}\right). \quad (3.18)$$

Letting $\theta = 0$, we again obtain the result that $P(T < \infty) = 1$, but the moment-generating function of T,

$$m_T(\theta) = \exp(-\sqrt{-2\lambda\theta}), \qquad \theta = 0, \tag{3.19}$$

is no longer that of an inverse Gaussian distribution but rather that of a stable distribution with index 1/2. See Feller (1966, p. 170). Of course, it is easy to see intuitively that with a drift coefficient of $v = $ zero, the inverse Gaussian mean becomes infinite and the density function should assume the form

$$f(x) = \frac{\lambda^{1/2}}{\sqrt{2\pi x^3}} \, e^{-\lambda/2x}.$$

3.2.5 Summary of Wiener Process with Drift

When the drift is positive, $P(T < \infty) = 1$ and the first passage time to the barrier is inverse Gaussian, $IG(\mu, \lambda)$, where

$$\mu = \frac{a - x_0}{v}$$

and

$$\lambda = \frac{(a - x_0)^2}{\sigma^2}.$$

When v is negative, $P(T < \infty) = e^{2\lambda/\mu}$ and the conditional distribution of T given $T < \infty$ is $IG(-\mu, \lambda)$. When $v = 0$, $P(T < \infty) = 1$ but the distribution of T is no longer inverse Gaussian but is the stable distribution with index 1/2.

Note that if we are given an inverse Gaussian distribution, we have no way of knowing whether it arose from a Wiener process with positive drift or from a Wiener process with a negative drift.

3.2.6 The Defective Inverse Gaussian Distribution

Whitmore (1979) discusses what he terms a defective inverse Gaussian distribution. He is referring to the family of *density functions*:

$$f(x; \mu, \lambda) = \sqrt{\frac{\lambda}{2\pi x^3}} \exp\left[-\frac{\lambda(x - \mu)^2}{2\mu^2 x} \right]$$

with $x > 0, \lambda > 0, -\infty < \mu < \infty, \mu \neq 0$. (3.20)

Of course, with $\mu < 0$, f does not integrate to unity so it is not a density function in the ordinary sense. Rather,

$$f(x; \mu, \lambda) = e^{-2\lambda/\mu} f(x; -\mu, \lambda).$$ (3.21)

3.3 WALD'S DISTRIBUTION: SAMPLE SIZE DISTRIBUTION IN A SEQUENTIAL PROBABILITY RATIO TEST

Wald (1944) considered the following sequential stopping rule for $S_N = Z_1 + Z_2 + \cdots + Z_N$, where $Z_i, i = 1, 2, \ldots$, are independent and identically distributed random variables with moment-generating function $\phi(t)$: stop if $S_N < b$ or if $S_N > a$. He obtained an approximate characteristic function for the random variable N by considering S_N to be exactly a or b upon stopping. In the same paper Wald obtained the characteristic function for N using the fundamental identity of sequential analysis that $E[e^{S_N t}(\phi(t))^{-N}] = 1$. He showed for the special case $b = -\infty$, $a = $ constant (>0), with $E(Z) > 0$ that the characteristic function of N is given by

$$\psi_N(\tau) = E(e^{\tau N}), \qquad \tau \text{ strictly imaginary}$$
$$= e^{at(\tau)}$$ (3.22)

where $t(\tau)$ is a root of the equation

$$-\log \phi(t) = \tau$$

such that $\lim_{\tau \to 0} t(\tau) = 0$.

When Z_i is normal with mean m and variance σ^2,

$$\log \phi(t) = mt + \frac{\sigma^2 t^2}{2}.$$

Then

$$t(\tau) = \frac{-m + \sqrt{m^2 - 2\sigma^2 \tau^2}}{\sigma^2}. \tag{3.23}$$

This gives the approximate characteristic function for N as

$$\psi_N(\tau) = \exp\left\{\frac{am}{\sigma^2}\left[1 - \left(1 - \frac{2\sigma^2}{m^2}\tau\right)^{1/2}\right]\right\}. \tag{3.24}$$

The probability density function $f(n)$ of N is obtained by inverting the characteristic function $\psi_N(t)$. As Wald showed, this results in the following form of $f(n)$

$$f(n) = \frac{a}{\sqrt{2\pi}}n^{-3/2}\exp\left[-\frac{m^2}{2n\sigma^2}\left(n - \frac{a}{m}\right)^2\right]. \tag{3.25}$$

One easily recognizes that $f(n)$ in (3.25) is the probability density function of an inverse Gaussian with parameters $\mu = a/m$ and $\lambda = a^2/\sigma^2$.

Of course, N is a discrete random variable so Wald's distribution must be viewed as an approximation. However, it seems reasonable on intuitive grounds. The distribution should be skewed to the right on the nonnegative half of the real line; $E[S_N]$ should be close to a. Also because S_N is the sum of N identically distributed random variables, each with mean m, $E(S_N)$ should be close to $mE(N)$. So we have

$$E(S_N) \doteq a$$

and

$$E(S_N) \doteq mE(N).$$

Therefore it seems reasonable that $E(N) \doteq a/m$.

Wald presented these results again in 1945 and in *Sequential Analysis* published in 1947. As pointed out by Johnson and Kotz

(1970) the inverse Gaussian distribution can be obtained as the approximate distribution of N under the more general situation where the Z_i are independent and identically distributed random variables with finite variance and positive mean.

3.4 INVERSION LAW: A FUNDAMENTAL CHARACTERISTIC OF INVERSELY RELATED VARIABLES

Tweedie (1945) observed the inverse relationship between the cumulant-generating functions of the following pairs of random variables:

    (a) Binomial and negative binomial
    (b) Poisson and gamma
    (c) Normal and inverse Gaussian

From the viewpoint of stochastic processes or sampling the two distributions in each case are intrinsically related to one another. In general, the relationship can be formulated as follows:

It is convenient to work with the logarithm of the Laplace transform $E[e^{-tZ}]$ of the random variable Z, which is in a sense a cumulant-generating function and will be so called in our presentation. Let

$$L_Z(t) = \log(E[e^{-tZ}]), \qquad t > 0.$$

Suppose $L_Z^{-1}(t)$ is an inverse function of $L_Z(t)$. We now define the inversely related variables as follows:

Definition Two random variables X and Y are said to be inverse variables if their cumulant-generating functions (cgf) $L_X(t)$ and $L_Y(t)$ satisfy the condition

$$L_X(t) = \alpha L(t)$$
$$L_Y(t) = \beta L^{-1}(t) \tag{3.26}$$

for all t values common to the domain of both cgf's and where α and

β are some constants and $L(L^{-1}(t)) = t$. Pairs of distributions having cgf's $\alpha L(t)$ and $\beta L^{-1}(t)$ will be called inverse distributions.

We shall now discuss L and L^{-1} for the distributions in (a), (b), and (c).

(a) Binomial and Negative Binomial

Consider the binomial distribution $B(n, p)$ with probability function

$$P(x) = \binom{n}{x} p^x (1-p)^{n-x}, \quad x = 0, 1, 2, \ldots, n, \ 0 \leqslant p \leqslant 1, q = 1 - p.$$

Then

$$L_B(t) = n \log(q + pe^{-t}), \qquad -\infty < t < \infty. \tag{3.27}$$

Let

$$L(t) = \log(q + pe^{-t}) \tag{3.28}$$

so

$$L_B(t) = nL(t).$$

Next by inverting $L(t)$ in (3.28), one obtains

$$L^{-1}(t) = \log[p(e^t - q)^{-1}], \qquad t > \log q. \tag{3.29}$$

Now consider the negative binomial distribution $NB(r, p)$ with probability function

$$P(y) = \binom{y-1}{r-1} p^r q^{y-r}, \qquad y = r, r+1, r+2, \ldots.$$

It is easily seen that

$$L_{\text{NB}}(t) = r \log[p(e^t - q)^{-1}], \qquad t > \log q. \tag{3.30}$$

Comparing (3.30) with (3.29), we have

$$L_{\mathrm{NB}}(t) = rL^{-1}(t).$$

Thus, when $L(t)$ is taken as in (3.28), we are able to show that

$$L_{\mathrm{B}}(t) = nL(t) \qquad \text{and} \qquad L_{\mathrm{NB}}(t) = rL^{-1}(t),$$

which satisfy the condition (3.26) with $\alpha = n$ and $\beta = r$.
   Next, if we consider

$$L(t) = \log p(e^t - q)^{-1}, \qquad t > \log q,$$

one easily gets

$$L^{-1}(t) = \log(q + pe^{-t}), \qquad -\infty < t < \infty.$$

Accordingly, $L_{\mathrm{NB}}(t) = rL(t)$ and $L_{\mathrm{B}}(t) = nL^{-1}(t)$. Hence the binomial and negative distributions are a pair of inverse distributions.
   It was noticed that α and β correspond to the respective parameters n and r of the two distributions. When $n = r = 1$, we get $L_{\mathrm{B}}(t) = L_{\mathrm{NB}}^{-1}(t)$ and $L_{\mathrm{NB}}(t) = L_{\mathrm{B}}^{-1}(t)$. Thus the exact inverse relationship holds for the pair of Bernoulli and geometric distributions and carries over to the binomial and negative binomial distributions according to our definition of inverse random variables.

(b)   Poisson and Gamma

Suppose that the number of occurrences for a certain type of event during any time interval of duration T is a Poisson variable with mean λT. For this Poisson distribution $P(\lambda T)$ the probability function

$$P(x) = \frac{e^{-\lambda T}(\lambda T)^x}{x!}, \qquad x = 0, 1, 2, \ldots,$$

and

$$L_p(t) = -\lambda T(1 - e^{-t}). \tag{3.31}$$

Consider

$$L(t) = -\lambda(1 - e^{-t}).$$

Then it is easy to see that the inverse function

$$L^{-1}(t) = \log\left(\frac{\lambda}{\lambda + t}\right), \qquad t > -\lambda. \tag{3.32}$$

Next consider the gamma distribution $G(a, \theta)$ with probability density function

$$f(y) = \frac{\theta^a}{\Gamma(a)} y^{a-1} e^{-\theta y}, \qquad y \geq 0, \theta > 0, a > 0$$

Then we have

$$E[e^{-tY}] = \left(\frac{\theta}{\theta + t}\right)^a.$$

So

$$L_G(t) = a \log\left(\frac{\theta}{\theta + t}\right), \qquad t > -\theta. \tag{3.33}$$

A comparison between (3.33) and (3.32) shows that $L_G(t)$ is similar to $L^{-1}(t)$ up to a multiple.

Thus, with $\theta = \lambda$, the cgf's in (3.31) and (3.33) are related as follows:

$$L_P(t) = TL(t) \text{ and } L_G(t) = aL^{-1}(t).$$

Similarly, if we consider

$$L(t) = \log\left(\frac{\lambda}{\lambda + t}\right)$$

it follows that

$$L^{-1}(t) = -\lambda(1 - e^{-t})$$

and thereby we again see that the cgf's $L_P(t)$ and $L_G(t)$ are inversely related as

$$L_P(t) = TL^{-1}(t) \qquad \text{and} \qquad L_G(t) = aL(t).$$

Therefore the Poisson and gamma distributions constitute a pair of inverse distributions.

When these relations are compared with the condition in (3.26), once again α and β correspond to the parameters T and a of the two distributions. Moreover, when $T = a = 1$, we have $L_P(t) = L_G^{-1}(t)$ and vice versa. Thus the Poisson (with unit time) and the negative exponential have the exact inverse relationship which then carries over to the pair of general Poisson and gamma distributions.

(c)   Normal and Inverse Gaussian

Consider the normal distribution $N(v, \sigma^2)$ with the probability density function

$$f(x) = \frac{1}{\sqrt{2\pi}\,\sigma} \exp\left[-\frac{(x - v)^2}{2\sigma^2} \right],$$

$$-\infty < x < \infty, \ -\infty < v < \infty, \ \sigma^2 > 0$$

Then

$$L_N(t) = -vt + \tfrac{1}{2}\sigma^2 t^2, \qquad -\infty < t < \infty. \tag{3.34}$$

Now suppose

$$L(t) = -t + \frac{\sigma^2}{2v}t^2, \qquad v \neq 0,$$

so that

$$L_N(t) = vL(t).$$

Restricting $L(t)$ to $t < 2v/\sigma^2$, we find the inverse of $L(t)$ to be

$$L^{-1}(t) = \frac{v}{\sigma^2}\left[1 - \left(1 + \frac{2\sigma^2}{v}t \right)^{1/2} \right]. \tag{3.35}$$

From Chapter 2 one is able to obtain the cgf of the inverse Gaussian. For the inverse Gaussian distribution with parameters μ and λ, we find that

$$L_{\text{IG}}(t) = \frac{\lambda}{\mu}\left[1 - \left(1 + \frac{2\mu^2}{\lambda}t\right)^{1/2}\right] \tag{3.36}$$

so it follows from equations (3.35) and (3.36) that

$$L_{\text{IG}}(t) = \mu L^{-1}(t)$$

when $\lambda/\mu^2 = v/\sigma^2$. Accordingly, we have $L_N(t) = vL(t)$ and $L_{\text{IG}}(t) = \mu L^{-1}(t)$. On the other hand, if we consider

$$L(t) = \frac{\lambda}{\mu^2}\left[1 - \left(1 + \frac{2\mu^2 t}{\lambda}\right)^{1/2}\right]$$

we are able to see that $L_N(t) = vL^{-1}(t)$ and $L_{\text{IG}}(t) = \mu L(t)$ with $\lambda/\mu^2 = v/\sigma^2$. Thus the normal and the inverse Gaussian are a pair of inverse distributions.

Once again, if $v = \mu = 1$, there is an exact relationship of $L_N(t) = L_{\text{IG}}^{-1}(t)$ and vice versa. This situation corresponds to the case of a fixed unit time or unit distance scale, whichever is applicable, in observing a Wiener process.

For the pair of inverse distributions as defined above, the following results hold.

Theorem 3.3 Let K_1 and K_2 be the first and second cumulants of the cgf $L(t)$. Then means and variances of the random variables X and Y satisfying (3.26) are given by

$$\mu_X = \alpha K_1, \qquad \sigma_X^2 = \alpha K_2$$
$$\mu_Y = \frac{\beta}{K_1}, \qquad \sigma_Y^2 = \frac{\beta K_2}{K_1^3} \tag{3.37}$$

and their coefficients of variation are equal if and only if $\beta = \alpha K_1$, in which case

$$\mu_X = \beta \qquad \text{and} \qquad \mu_Y = \alpha. \tag{3.38}$$

Proof: Since $L_X(t) = \alpha L(t)$ and K_1 and K_2 are the mean and variance of the random variable having cgf $L(t)$, the results are obvious for the random variable X. In the case of random variable Y, the results follow because

$$\frac{d}{dt} L^{-1}(t) = [L'(t)]^{-1}$$

$$\frac{d^2}{dt^2} L^{-1}(t) = L''(t)[L'(t)]^{-3}$$

where L' and L'' denote the first and second derivatives of L with respect to t.

The coefficients of variation of X and Y are $CV_X = \sqrt{K_2}/\sqrt{\alpha K_1}$ and $CV_Y = \sqrt{K_2}/\sqrt{\beta K_1}$. So $CV_X = CV_Y$ if and only if $\alpha K_1 = \beta$. Thus (3.38) follows.

4

Certain Useful Transformations
and Characterizations

4.1 RELATED STATISTICAL DISTRIBUTIONS

We give here certain distributions related to the inverse Gaussian. The distributions have a close analogy with those related to the normal distribution, for example, Student's t, chi-square, and F. These distributions, among others, are given by Chhikara (1972) and Chhikara and Folks (1975).

We previously obtained in Theorem 2.1 a distribution which relates the inverse Gaussian to the normal. For $X \sim \text{IG}(\mu, \lambda)$ the transformed variable $Y = \sqrt{\lambda}(X-\mu)/\mu\sqrt{X}$ is distributed according to the density function in (2.15). We call the distribution of Y a nonlinear weighted normal because its density function is that of the standard normal multiplied by a nonlinear function of y.

Theorem 4.1 $|Y|$ is distributed as standard half-normal.

Proof: Let $W = |Y|$. $F_W(w) = P(W \leqslant w) =$

$$P(-w \leqslant Y \leqslant w) = \int_{-w}^{w} (1 + h(y)) \frac{1}{\sqrt{2\pi}} e^{-y^2/2} \, dy$$

where $h(y)$ is an odd function. So differentiating, we obtain

$$f_W(w) = \frac{2}{\sqrt{2\pi}} e^{-w^2/2}.$$

Theorem 4.2 Y^2 has the chi-square distribution with one degree of freedom.

Proof: Note that $Y^2 = W^2$ with W as given in Theorem 4.1. Then we obtain immediately that Y^2 is a chi-square variable with one degree of freedom.

4.2 ANALOGY WITH THE NORMAL DISTRIBUTION THEORY

In this section we summarize several of the results given by Chhikara and Folks (1975). First we give a more general result than in Theorem 2.1 referred to in the preceding section.

Theorem 4.3 Let $X \sim \text{IG}(\mu, \lambda)$. Then the density function of $T = \sqrt{\lambda}(X - \mu_0)/\mu_0\sqrt{X}$, where $\mu_0 > 0$, is given by

$$p(t; \mu, \lambda) = \frac{1}{\sqrt{2\pi}}\left[\exp\left(-\frac{\lambda}{2\mu_0}\right)\left(1 - \frac{\mu_0}{\mu}\right)^2\right]\left(1 - \frac{t}{\sqrt{4\lambda/\mu_0 + t^2}}\right)$$

$$\times \exp\left[-\frac{t^2}{4}\left(1 + \frac{\mu_0^2}{\mu^2}\right)\right.$$

$$\left. + \frac{t\sqrt{4\lambda/\mu_0 + t^2}}{4}\left(1 - \frac{\mu_0^2}{\mu^2}\right)\right], \qquad \infty < t < \infty. \quad (4.1)$$

This result is analogous to the distribution of $(X - \mu_0)/\sigma$ where $X \sim N(\mu, \sigma^2)$. The density function of $|T|$ can be obtained from (4.1) but it does not enjoy the folding property as in the case with the distribution of Y in Section 4.1.

Theorem 4.4 Let $X \sim \text{IG}(\mu, \lambda)$ and $Z \sim \chi^2(n)/\lambda$ be two independent random variables. Then the density function of

$$U = \frac{\sqrt{n}(X - \mu)}{\mu\sqrt{XZ}}$$

is

$$h(u; \lambda/\mu) = [1 - c(u)]f_{t,n}(u), \qquad -\infty < u < \infty \qquad (4.2)$$

where

$$c(u) = \frac{u}{\Gamma[(n + 1)/2]} \int_0^\infty \frac{y^{n/2}e^{-y}}{\sqrt{2(n + u^2)\lambda/\mu + u^2 y}} \, dy \qquad (4.3)$$

and $f_{t,n}(u)$ is the density function of Student's t with n df.

The distribution of U is unimodal but unsymmetric. Note that $c(u)$ is an odd function in u. We call this distribution of U a nonlinear weighted Student's t. It follows as a corollary that $|U|$ is a truncated Student's t variable on the positive real line and that U^2 has the F distribution with 1 and n degrees of freedom.

Although not quite as striking in appearance, a distribution analogous to the noncentral Student's t is given in the following theorem.

Theorem 4.5 Let $X \sim IG(\mu, \lambda)$ be independent of random variable $Z \sim \chi^2(n)/\lambda$. Suppose $W = \sqrt{n}(X - \mu_0)/\mu_0\sqrt{XZ}$, $\mu_0 > 0$. Then the density function of W is as follows.

$$s(w; \mu, \lambda) = \frac{\exp[-\lambda(\mu - \mu_0)^2/2\mu^2\mu_0]}{\sqrt{n\pi}\,\Gamma(n/2)} \left(2 + \frac{\mu^2 + \mu_0^2}{\mu^2} \frac{w^2}{n}\right)^{-(n+1)/2}$$

$$\times \sum_{k=0}^\infty \frac{1}{k!} \left[\frac{w/\sqrt{n}}{\{2 + [(\mu^2 + \mu_0^2)/\mu^2]w^2/n\}}\right]^k$$

$$\times \left[\frac{\mu^2 - \mu_0^2}{2\mu^2\mu_0}\right]^k \left[G_{n+2k-1,k}(w) - \frac{\mu_0 w}{\sqrt{n}} G_{n+2k-1,k-1}(w)\right],$$

$$-\infty < w < \infty \qquad (4.4)$$

where

$$G_{ij}(w) = \int_0^\infty \left[2\lambda\mu_0\left(2 + \frac{\mu^2 + \mu_0^2}{\mu^2}\frac{w^2}{n}\right) + \frac{\mu_0^2 w^2}{n}u\right]^{j/2} u^{(i-j)/2}e^{-u/2} \, du.$$

Next we give a nonlinear weighted noncentral chi-square distribution.

Theorem 4.6 Let $X \sim \text{IG}(\mu, \lambda)$. Then the distribution of $T^2 = \lambda(X - \mu_0)^2/\mu_0^2 X$, $\mu_0 > 0$, is given by the density function

$$q(t^2; \mu, \lambda) = \exp\left[\frac{-\lambda(\mu-\mu_0)^2}{2\mu^2\mu_0}\right] \sum_{n=0}^{\infty} \frac{[(\mu^2-\mu_0^2)/8\mu^2]^{2n}(t^2)^{n-1/2}}{\sqrt{2}n!\mu_0^n\Gamma(n+1/2)}$$

$$\times \left[1 - \frac{(\mu^2-\mu_0^2)t^2}{4\mu^2(2n+1)}\right](4\lambda + \mu_0 t^2)^n$$

$$\times \exp\left[\frac{-(\mu^2+\mu_0^2)t^2}{4\mu^2}\right], \qquad t^2 > 0. \qquad (4.5)$$

When $\mu = \mu_0$ in (4.5), one obtains the central χ^2 density function with one degree of freedom. Otherwise the density function which we call the nonlinear weighted noncentral χ^2 results.

Not only do the results in this section point out the strong analogy that exists between the inverse Gaussian and the normal distributions; these results are very helpful in developing statistical methodology based on the inverse Gaussian distribution and they are used in Chapters 5 and 6.

For further discussion on these related statistical distributions, refer to Chhikara and Folks (1975) and Folks and Chhikara (1978). Later, in Section 4.4, we give several results which outline further the nature of the analogy that exists between the IG and normal distribution.

4.3 DISTRIBUTION OF THE RECIPROCAL OF THE INVERSE GAUSSIAN VARIABLE

Sometimes it is convenient to work with the reciprocal of the inverse Gaussian variable. For example, the maximum likelihood estimates of IG parameters given in the next chapter and the analysis of residuals in Chapter 6 can be expressed simply in terms of this variable. It also holds promise for a development of linear regression analysis for the inverse Gaussian. Thus the distribution and useful properties of this variable merit a short discussion. For further details, refer to Tweedie (1957a, b).

When $X \sim IG(\mu, \lambda)$, the density function f' of $W = X^{-1}$ is given by

$$f'(w; \mu, \lambda) = \left(\frac{\lambda}{2\pi w}\right)^{1/2} \exp\left[\frac{-\lambda(1 - \mu w)^2}{2\mu^2 w}\right], \qquad w > 0. \qquad (4.6)$$

Equivalently,

$$f'(w; \mu, \lambda) = \mu w f(w; \mu^{-1}, \lambda \mu^{-2}), \qquad (4.7)$$

where f is of the form given in (2.1).

The density function of W is unimodal and positively skewed. The mode occurs at

$$\frac{1}{\mu}\left[\left(1 + \frac{\mu^2}{4\lambda^2}\right)^{1/2} - \frac{\mu}{2\lambda}\right].$$

Its density curves are given for $\mu = 1$ and $\lambda = 0.2, 0.5, 1, 5, 10$, and 30 in Figure 4.1 and for $\lambda = 1$ and $\mu = 1, 2, 3$ and 4 in Figure 4.2. When these curves are compared with their counterparts in Figures 2.1 and 2.2 the difference is rather striking for small values of λ considered in Figures 2.1 and 4.1, and for the largest value of μ in Figures 2.2 and 4.2.

All moments of W exist. Its positive moments can be easily obtained from those of X because of (2.11). That is,

$$E[W^r] = \frac{E[X^{r+1}]}{\mu^{2r+1}}. \qquad (4.8)$$

It follows from the characteristic function of X given in (2.4) that

$$E(W^r) = \mu^{-r} \sum_{s=0}^{r} \frac{(r+s)!}{s!(r-s)!}\left(\frac{2\lambda}{\mu}\right)^{-s}. \qquad (4.9)$$

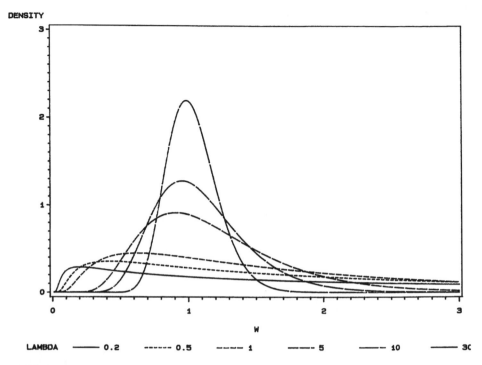

Figure 4.1 Probability densities for IG reciprocal with $\mu = 1$ for six values of λ.

In particular,

$$E(W) = \frac{1}{\mu} + \frac{1}{\lambda}$$

$$Var(W) = \frac{1}{\lambda\mu} + \frac{2}{\lambda^2}, \tag{4.10}$$

and the measures of skewness and kurtosis are

$$\frac{3\phi + 8}{(\phi + 2)^{3/2}} \tag{4.11}$$

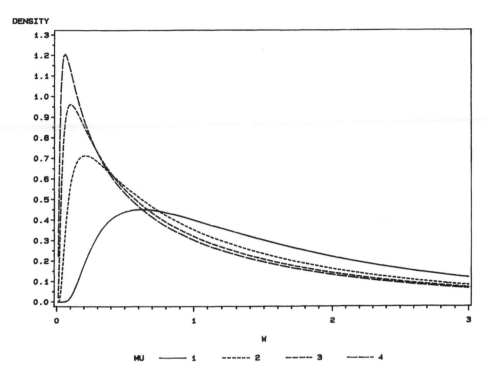

Figure 4.2 Probability densities for IG reciprocal with $\lambda = 1$ for four values of μ.

and

$$\frac{3(5\phi + 16)}{(\phi + 2)^2},$$

respectively; again $\phi = \lambda/\mu$. Naturally, both skewness and kurtosis depend on the shape parameter ϕ, but the expressions in (4.11) are not as simple as for the variable X given in (2.10).

Next, the cumulant-generating function of W,

$$\log E[e^{-tw}] = \log \int_0^\infty e^{tw} f'(w; \mu, \lambda)\, dw$$

$$= \log[e^{(\lambda/\mu)[1-(1+2t/\lambda)]^{1/2}}]$$

$$\times \int_0^\infty f'\left[\left(w\left(1+\frac{2t}{\lambda}\right)^{1/2}; \mu, \lambda\left(1+\frac{2t}{\lambda}\right)^{1/2}\right) dw\right]$$

$$= \log e^{(\lambda/\mu)[1-(1+2t/\lambda)^{1/2}]} \frac{1}{(1+2t/\lambda)^{1/2}}$$

$$= \frac{\lambda}{\mu}\left[1-\left(1+\frac{2t}{\lambda}\right)^{1/2}\right] - \frac{1}{2}\log\left(1+\frac{2t}{\lambda}\right). \qquad (4.12)$$

The formula (4.12) shows that the random variable,

$$W \sim \mathrm{IG}(\mu^{-1}, \lambda\mu^{-2}) \circ \chi_1^2/\lambda,$$

that is, the distribution of W is a convolution of an inverse Gaussian distribution with an independent distribution of a chi-square variable (multiplied by λ^{-1}) with one degree of freedom. This characteristic of W makes its distributional and inferential properties tractable, as will be seen later. Moreover, Kirmani and Ahsanullah (1986) gave a characterization of the inverse Gaussian distribution based on this property of random variable W. Also, see Ahsanullah and Kirmani (1984) for a characterization of the Wald distribution.

4.4   SOME CHARACTERIZATIONS OF THE INVERSE GAUSSIAN DISTRIBUTION

First we present a characterization of the inverse Gaussian distribution which is analogous to the characterization of the normal distribution based on the independence of sample mean and variance. This result is due to both Tweedie (1957a) and Khatri (1962). The following theorem states the results as established by them in two parts.

Theorem 4.7   Let $X_1, X_2, \ldots, X_n$ be independently and identically distributed positive random variables such that the expected values of $X, X^2, 1/X$, and $1/\Sigma X_i$ exist and are different from zero. Then the necessary and sufficient condition that $X_1, X_2, \ldots, X_n$ are distributed as inverse Gaussian is that ΣX_i and $\Sigma X_i^{-1} - n^2(\Sigma X_i)^{-1}$ are independently distributed.

Proof: (NECESSITY) This is the proof given by Tweedie. Let X_1, $X_2, \ldots, X_n$ be iid $IG(\mu, \lambda)$, $\bar{X} = \Sigma X_i/n$, and $V = \Sigma (1/X_i - 1/\bar{X})$. We now show that the conditional distribution of V given $\bar{X}$ is independent of $\bar{X}$. To do this we shall find the conditional moment-generating function of V given $\bar{X} = \bar{x}$. This can be found by taking expectation with respect to the conditional density function of X_1, $X_2, \ldots, X_n$ given $\bar{X} = \bar{x}$.

More properly, since there is a functional relationship between $X_1, X_2, \ldots, X_n$ and $\bar{X}$, we can take expectation with respect to X_1, X_2, $X_3, \ldots, X_{n-1}$ given $\bar{X} = \bar{x}$. Let $Y_1 = X_1$, $Y_2 = X_2, \ldots$, $Y_{n-1} = X_{n-1}$, $Y_n = \bar{X}$. The Jacobian of this transformation is n. Then the joint density function of $Y_1, Y_2, \ldots, Y_n$ is

$$f(y_1, y_2, \ldots, y_n) = n \prod_{i=1}^{n} \left[\frac{\lambda}{2\pi x_i^3} \right]^{1/2} \exp\left[-\frac{\lambda}{2\mu^2} \sum_{i=1}^{n} \frac{(x_i - \mu)^2}{x_i} \right],$$

(4.13)

where the x's are expressed in terms of the y's. Because of the factorization

$$\frac{\lambda}{2\mu^2} \sum \frac{(x_i - \mu)^2}{x_i} = \frac{\lambda}{2} \sum \left(\frac{1}{x_i} - \frac{1}{\bar{x}} \right) + \frac{n\lambda(\bar{x} - \mu)^2}{2\mu^2 \bar{x}},$$

(4.14)

we can write the right-hand side of (4.13) as

$$n \prod_{i=1}^{n} \left(\frac{\lambda}{2\pi x_i^3} \right)^{1/2} \exp\left[-\frac{\lambda v}{2} - \frac{n\lambda(\bar{x} - \mu)^2}{2\mu^2 \bar{x}} \right].$$

(4.15)

It follows from the additive property discussed in Section 2.4 that $\bar{X} = Y_n \sim I(\mu, n\lambda)$. Thus the conditional density of $Y_1, Y_2, \ldots, Y_{n-1}$ given Y_n is given by

$$f(y_1, y_2, \ldots, y_{n-1} | y_n) = \sqrt{n\bar{x}^3} \prod_{i=1}^{n-1} \left(\frac{\lambda}{2\pi} \right)^{1/2} \prod_{1}^{n} x_i^{-3/2} e^{-(\lambda v/2)}$$

(4.16)

where the x's and v are expressed in terms of the y's.

Now consider the conditional moment-generating function of V given $\bar{X} = \bar{x}$.

$$E(e^{tV}) = \int_{y_1=0}^{\infty} \int_{y_2=0}^{\infty} \cdots \int_{y_{n-1}=0}^{\infty} e^{tv} \sqrt{n\bar{x}^3} \prod_{i=1}^{n-1} \left(\frac{\lambda}{2\pi}\right)^{1/2}$$

$$\times \prod_{i=1}^{n} x_i^{-3/2} e^{-(\lambda/2)v} \prod_{i=1}^{n-1} dx_i$$

$$= \left(1 - \frac{2t}{\lambda}\right)^{-(n-1)/2} \int_{y_1=0}^{\infty} \int_{y_2=0}^{\infty} \cdots \int_{y_{n-1}=0}^{\infty} \sqrt{n\bar{x}^3}$$

$$\times \prod_{i=1}^{n-1} \left[\frac{\lambda[1(-2t/\lambda)]}{2\pi}\right]^{1/2} \prod_{i=1}^{n} x_i^{-(3/2)}$$

$$\times \exp\left[-\frac{\lambda}{2}\left(1 - \frac{2t}{\lambda}\right)v\right] \prod_{i=1}^{n-1} dx_i$$

$$= \left(1 - \frac{2t}{\lambda}\right)^{-(n-1)/2} \tag{4.17}$$

because the integral in (4.17) is just the integral of the density function in (4.16) with λ replaced by $\lambda(1 - 2t/\lambda)$. Because the conditional moment-generating function of V is the same for all $\bar{x}$, we conclude that V is independent of $\bar{X}$, or ΣX_i. Also it follows that $\lambda V \sim \chi_{n-1}^2$.

(SUFFICIENCY)   We here simply outline major steps of the proof given by Khatri (1962). Those interested in further details may refer to his paper. Let V be as before and $S = \Sigma X_i$.

Let $\phi(t)$ and $F(x)$ be the characteristic function and the distribution function of random variable X. Then under the conditions of Theorem 4.7, it easily follows that

$$\int x^{-1} e^{itx} \, dF(x) = i \int_{-T}^{t} \phi(p) \, dp + \int x^{-1} e^{-iTx} \, dF(x) \tag{4.18}$$

and

$$\int \cdots \int s^{-1} e^{its} \prod_{j=1}^{n} dF(x_j) = i \int_{-T}^{t} [\phi(p)]^n \, dp$$

$$+ \int \cdots \int s^{-1}e^{-iTs} \prod_{j=1}^{n} dF(x_j). \qquad (4.19)$$

Then

$$E[Ve^{itS}] = E\left[\sum_{j=1}^{n} X_j^{-1}\left(\prod_{k=1}^{n} e^{itX_k} \right) \right] - n^2 E[S^{-1}e^{itS}] \qquad (4.20)$$

can be expressed in terms of the right sides of equations (4.18) and (4.19). On the other hand, the independence of V and S implies that

$$E[Ve^{itS}] = E(V)E[e^{itS}] = (n-1)\delta[\phi(t)]^n \qquad (4.21)$$

by letting $E[V] = (n-1)\delta$. Equating the resulting expression from (4.21) to $(n-1)\delta[\phi(t)]^n$ and differentiating it twice with respect to t, it can be shown that

$$\phi(t) = i\delta\phi'(t) + \{[\phi(t)]^2\phi''(t)/[\phi'(t)]^2\} \qquad (4.22)$$

for all real values of t for which $\phi(t) \neq 0$ and $\phi'(t) \neq 0$. Here ϕ' and ϕ'' denote the first and second derivatives of ϕ.

The differential equation in (4.22) can be easily solved by letting $h(t) = \log \phi(t)$ and then seeing that the equation in (4.22) is equivalent to

$$h''(t)[h'(t)]^{-3} = -i\delta. \qquad (4.23)$$

Integrating (4.23) and letting $h'(0) = i\xi$, we have

$$[h'(t)]^{-2} = \frac{-(1 - 2i\delta\xi^2 t)}{\xi^2}.$$

Now taking the square root of its inverse and then integrating the resulting expression, it follows that

$$h(t) = \pm \frac{[1 - (1 - 2i\delta\xi^2 t)^{1/2}]}{\delta\xi}$$

and then $\phi(t)$ is given by $\exp[h(t)]$. This is true for all real values of t for which $\phi(t) \neq 0$ and $\phi'(t) \neq 0$. But due to the uniqueness property of the characteristic function, the random variable X is distributed as inverse Gaussian with parameters $\pm \xi$ and δ^{-1}. Hence X_1, $X_2, \ldots, X_n$ are inverse Gaussian distributed. This completes the proof.

Seshadri (1983) noticed that Khatri used independence of V and S only to justify Equation (4.21). Now the condition that

$$E[Ve^{itS}] = E(V)E[e^{itS}]$$

is shown by Kagan, Linnik, and Rao (1973) to hold if and only if V has constant regression on S, that is, if and only if

$$E[V|S] = E[V].$$

Accordingly, Seshadri weakened the condition of independence in Khatri's characterization to give the following theorem.

Theorem 4.8 Let X_1, $X_2, \ldots, X_n$ be independent, identically distributed random variables such that $E(X_i)$, $E(X_i^2)$, $E(X_i^{-1})$, and $E(\Sigma_1^n X_i)^{-1}$ exist and are nonzero. If the regression of V on S is constant, then each X_i has an inverse Gaussian distribution.

Of course, as Seshadri points out, the constancy of regression is both necessary and sufficient.

Letac and Seshadri (1983) relax the conditions on existence of moments in Theorem 4.8 to a condition of positivity and show that independence of V and S is sufficient to conclude that X_i has either the inverse Gaussian or stable distribution. We state this in the following theorem.

Theorem 4.9 Let $X_1, X_2, \ldots, X_n$ $(n \geqslant 2)$ be independent identically distributed positive nondegenerate random variables. Then V and S are independent only if the density of X_i is either inverse Gaussian or stable.

Letac and Seshadri (1983) give a characterization of the generalized inverse Gaussian distribution described briefly in Section

2.8. Using their parameterization, the density function for a generalized inverse Gaussian distribution is denoted by

$$f_{\lambda,a,b}(x) = \frac{a^{\lambda/2}b^{-\lambda/2}}{2K_\lambda(\sqrt{ab})} x^{\lambda-1} \exp[-\tfrac{1}{2}(ax + bx^{-1})] \qquad (4.24)$$

where $-\infty < \lambda < \infty$ and $a, b, x > 0$. A gamma density function is denoted by

$$g_{\lambda,a}(x) = \frac{a^{-\lambda}}{\Gamma(\lambda)} x^{\lambda-1} \exp(-a^{-1}x) \qquad (4.25)$$

where $\lambda, a > 0$.

Now their characterization can be stated.

Theorem 4.10 Let $X > 0$ be independent of $Y \sim g_{\lambda,2/a}$, $\lambda, a > 0$. Then X has the same distribution as $1/(Y + X)$ if and only if $X \sim f_{-\lambda,a,a}$.

Of course, with suitable choice of parameters, the process can be continued to characterize the distribution of X by equivalence of the variables

$$X \quad \text{and} \quad \frac{1}{Y + X},$$

$$X \quad \text{and} \quad \frac{1}{Y_1 + 1/(Y_2 + X)},$$

$$X \quad \text{and} \quad \frac{1}{Y_1 + \dfrac{1}{Y_2 + 1/(Y_3 + X)}},$$

and so on.

In the necessity proof of Theorem 4.7 it was obtained that $\lambda V \sim X^2_{n-1}$. Because of the strong analogy with the normal distribution, a natural question is whether λV can be decomposed into the sum of independent chi-square variables, each with a single degree of freedom. Seshadri (1981, 1983) gives such a decomposition where $n = 2^r$, r being a positive integer. To illustrate the idea, when $n = 4$, let

$$V_1 = \frac{1}{X_1} + \frac{1}{X_2} - \frac{4}{X_1 + X_2}$$

$$V_2 = \frac{1}{X_3} + \frac{1}{X_4} - \frac{4}{X_3 + X_4}$$

$$V_3 = \frac{4}{X_1 + X_2} + \frac{4}{X_3 + X_4} - \frac{16}{X_1 + X_2 + X_3 + X_4}$$

Then $V = V_1 + V_2 + V_3$ and λV_1, λV_2, and λV_3 are independent chi-square variables, each with one degree of freedom.

Another characterization is given by Roy and Wasan (1969) in terms of cubic regression. They show that the inverse Gaussian distribution is characterized by $E(Q \mid \Sigma X)$ being a cubic function of ΣX, where Q is a cubic function of the X_i's.

4.5 GENERATING RANDOM VARIATES FROM THE INVERSE GAUSSIAN DISTRIBUTION

A standard technique to generate random observations from a distribution is by inverting its cumulative distribution function (cdf). If the inverted cdf has a simple closed-form expression, its evaluations at random values from an acceptable uniform $(0, 1)$ generator provide an efficient method of getting random variates from the distribution. Unfortunately, the cdf of the inverse Gaussian given in (2.14) cannot be inverted easily; thus the preceding technique is not directly applicable.

Michael et al. (1976) gave a method of generating random variates using a transformation with multiple roots. Their basic approach is to find a transformation of the random variable of interest, and then to use the multinominal probabilities associated with the multiple roots of the transformation to choose one root for the random observation.

For $X \sim IG(\mu, \lambda)$, the transformed variable

$$Y^2 = \frac{\lambda(X - \mu)^2}{\mu^2 X}$$

distributed as χ_1^2 has two roots, say X_1 and X_2, where

$$X_1 = \frac{\mu}{2\lambda}[2\lambda + \mu Y^2 - \sqrt{4\lambda\mu Y^2 + \mu^2 Y^4}] \qquad (4.26)$$

and

$$X_2 = \frac{\mu^2}{X_1}. \qquad (4.28)$$

Given a random value Y^2 generated from the χ_1^2, Michael et al. (1976) gave the conditional probability with which each root should be selected; the smaller root X_1 should be chosen with probability $\mu/(\mu + X_1)$ and the other root with probability $X_1/(\mu + X_1)$. The overall procedure for generating inverse Gaussian variates is as follows:

1. Generate random numbers from the chi-square distribution with 1 degree of freedom.
2. For each random value in step 1, compute the smaller root X_1 given above.
3. Perform a Bernoulli trial with probability of "success" $p = \mu/(\mu + X_1)$.
4. If the trial results in a success, the root X_1 is chosen for the random observation from the inverse Gaussian distribution; otherwise the larger root X_2 is chosen.

Several other writers have considered generating inverse Gaussian variables. See, for example, Aitkinson (1979), who considered the simulation of generalized inverse Gaussian variables.

5

Sampling and Estimation of Parameters

5.1   MAXIMUM LIKELIHOOD ESTIMATORS

For a random sample $X_1, X_2, \ldots, X_n$ from an inverse Gaussian population $IG(\mu, \lambda)$, the likelihood function is

$$L = \left(\frac{\lambda}{2\pi}\right)^{n/2} \left(\prod_1^n x_i^{-3/2}\right) \exp\left[-\lambda \sum_1^n \frac{(x_i - \mu)^2}{2\mu^2 x_i}\right], \qquad \mu > 0, \lambda > 0$$

(5.1)

and the maximum likelihood estimators (MLEs) of μ and λ are

$$\hat{\mu} = \bar{X}$$
$$\frac{1}{\hat{\lambda}} = \frac{1}{n}\sum_1^n \left(\frac{1}{X_i} - \frac{1}{\bar{X}}\right)$$

(5.2)

where

$$\bar{X} = \frac{1}{n}\sum_1^n X_i$$

Suppose $\bar{X}_H$ is the sample harmonic mean so that

$$\bar{X}_H = 1 \Bigg/ \left[\frac{1}{n} \sum_1^n \frac{1}{X_i} \right]$$

Then $1/\hat{\lambda}$ in (5.2) can be rewritten as the difference $(1/\bar{X}_H - 1/\bar{X})$. Since $\bar{X} \geqslant \bar{X}_H$, the estimator $1/\hat{\lambda}$ is nonnegative. The estimators in (5.2) were obtained by Schrödinger (1915).

The estimators of μ and λ given in (5.2) are highly intuitive. It is natural to consider the sample average $\bar{X}$ to estimate the population mean μ. Next, the estimator of $1/\lambda$ given in (5.2) is very intuitive as well because from the result in (4.10) we have

$$\frac{1}{\lambda} = E \left[\frac{1}{X} \right] - \frac{1}{\mu}$$

It is natural to consider the sample average, $(1/n)\sum(1/X_i)$, for an estimate of $E[1/X]$ and $1/\bar{X}$ for an estimate of $1/\mu$; hence, a natural choice of estimator for $1/\lambda$ is the statistic $(1/\bar{X}_H - 1/\bar{X})$.

As discussed previously in Section 4.4, $\bar{X}$ and $V = \sum(1/X_i - 1/\bar{X})$ are independently distributed so that $\hat{\mu} \sim \mathrm{IG}(\mu, n\lambda)$ and $n\lambda/\hat{\lambda} \sim \chi^2_{n-1}$. The independence of the two statistics $\bar{X}$ and V also follows from the reproductive exponential family property of the inverse Gaussian distribution, as discussed by Barndorff-Nielsen and Blaesild (1983b). It is easily seen that $(\bar{X}, \bar{X}_H)$ is the minimal sufficient statistic for (μ, λ) and because the inverse Gaussian family is an exponential family, $(\bar{X}, \bar{X}_H)$ is complete. The statistic $(\sum X_i, \sum(1/X_i - 1/\bar{X}))$ is a one-to-one function of $(\bar{X}, \bar{X}_H)$ and is also a complete sufficient statistic for (μ, λ). The uniform minimum variance unbiased estimators (UMVUEs) of μ and $1/\lambda$ are $\bar{X}$ and $\sum_1^n(1/X_i - 1/\bar{X})/(n - 1)$, respectively.

In terms of the analogy with the normal distribution, the inferences concerning $1/\lambda$, as discussed in the next chapter, parallel those concerning the variance for the normal.

Moment estimators of μ and λ and their asymptotic variances were given by Patel (1965) for the one-sided and two-sided truncated inverse Gaussian distributions. His results are well summarized in Johnson and Kotz (1970).

5.2 VARIANCE ESTIMATION

The MLE of variance μ^3/λ is $\hat{\mu}^3/\hat{\lambda}$ where $\hat{\mu}$ and $1/\hat{\lambda}$ are given by (5.2). This estimate of variance is biased. To simplify the discussion regarding unbiased estimation of μ^3/λ consider the modified MLE of μ^3/λ given by

$$\frac{n}{n-1}\left(\frac{\hat{\mu}^3}{\hat{\lambda}}\right) \tag{5.3}$$

The bias of the modified MLE in (5.3) is

$$\frac{3\mu^4}{n\lambda^2}\left(1 + \frac{\mu}{n\lambda}\right) \tag{5.4}$$

and its mean square error is

$$\frac{1}{n-1}\left(\frac{\mu^3}{\lambda}\right)\left[\left(2 + 9\frac{\mu}{\lambda}\right) + \frac{3}{n}\frac{\mu}{\lambda}\left(7 + 33\frac{\mu}{\lambda}\right)\right] + O(n^{-3}). \tag{5.5}$$

The form of bias in (5.4) allows us to obtain by jackknifing (Gray and Schucany, 1972) an unbiased estimator of μ^3/λ given by

$$\left(\frac{n^2}{2}\right)\left(\frac{\hat{\mu}^3}{\lambda^*}\right) - (n-1)^2\left(\frac{\hat{\mu}^3}{\lambda^*}\right)^{(1)} + \left(\frac{n-2}{2}\right)^2\left(\frac{\hat{\mu}^3}{\lambda^*}\right)^{(2)} \tag{5.6}$$

where $\lambda^* = (n-1)\hat{\lambda}/n$, $(\hat{\mu}^3/\lambda^*)^{(1)}$ is the average of estimates $(\hat{\mu}^3/\lambda^*)_j$, $j = 1, 2, \ldots, n$, obtained by deleting one observation at a time and $(\hat{\mu}^3/\lambda^*)^{(2)}$ is the average of estimates $(\hat{\mu}^3/\lambda^*)_{i,j}$ where $i, j = 1, 2, \ldots, n, i \neq j$, obtained by deleting a pair of observations at a time.

The UMVUE of μ^3/λ was obtained by Korwar (1980) by applying the *Rao-Blackwell theorem*: given a complete sufficient statistic T for θ and an unbiased estimate $\tilde{g}(\theta)$ of a parametric function $g(\theta)$, the UMVUE of $g(\theta)$ is given by $\hat{g}(\theta) = E[\tilde{g}(\theta) | T]$. The sample variance

$$S^2 = \sum_{i=1}^{n} \frac{(X_i - \bar{X})^2}{n-1}$$

is an unbiased estimator for the variance and can be used to obtain the UMVUE. The statistic

$$T = (\bar{X}, V), \qquad \text{where } V = \sum_{1}^{n} \left(\frac{1}{X_i} - \frac{1}{\bar{X}} \right),$$

is a complete sufficient statistic for (μ, λ) and

$$E[(n-1)S^2 \,|\, T = (\bar{x}, v)] = nE[X_1^2 \,|\, \bar{x}, v] - n\bar{x}^2. \tag{5.7}$$

Chhikara and Folks (1974) derived the conditional distribution of X_1, given $T = (\bar{x}, v)$; the conditional density function is given by

$$h(x_1 \,|\, \bar{x}, v) = \frac{\sqrt{n(n-1)}}{B[\frac{1}{2}, (n-2)/2]} \left[\frac{\bar{x}^3}{vx_1^3(n\bar{x} - x_1)^3} \right]^{1/2}$$

$$\times \left[1 - \frac{n(x_1 - \bar{x}^2)}{vx_1\bar{x}(n\bar{x} - x_1)} \right]^{(n-4)/2}, \quad L < x_1 < U \tag{5.8}$$

where L and U are the smaller and larger roots of the equation

$$n(x_1 - \bar{x})^2 = vx_1\bar{x}(n\bar{x} - x_1).$$

It can be shown that these roots are as follows:

$$L = \bar{x} \, \frac{n[2 + v\bar{x}] - [4n(n-1)v\bar{x} + n^2v^2\bar{x}^2]^{1/2}}{2[n + v\bar{x}]}$$

$$U = \bar{x} \, \frac{n[2 + v\bar{x}] + [4n(n-1)v\bar{x} + n^2v^2\bar{x}^2]^{1/2}}{2[n + v\bar{x}]}.$$

Thus

$$E[X_1^2 \,|\, \bar{x}, v] = \int_{L}^{U} x_1^2 h(x_1 \,|\, \bar{x}, v) \, dx_1$$

$$= \frac{\sqrt{n(n-1)}}{B[\frac{1}{2}, (n-2)/2]} \int_{L}^{U} \left[\frac{x_1 \bar{x}^3}{v(n\bar{x} - x_1)^3} \right]^{1/2}$$

$$\times \left[1 - \frac{n(x_1 - \bar{x})^2}{vx_1\bar{x}(n\bar{x} - x_1)} \right]^{(n-4)/2} dx_1. \tag{5.9}$$

The integrand on the right side in (5.9) is a complex function of x_1 and it is not easy to get a closed-form expression for the right side. One may however simplify it by considering the approach given in Chhikara and Folks (1974) based on the following transformation:

$$w = \frac{\sqrt{n}(x_1 - \bar{x})}{\sqrt{vx_1\bar{x}(n\bar{x} - x_1)}}\left[1 - \frac{n(x_1 - \bar{x})^2}{vx_1\bar{x}(n\bar{x} - x_1)}\right]^{-1/2}.$$

It can be shown that the integral in (5.9) simplifies to

$$\frac{\bar{x}^2}{B[\frac{1}{2}, (n-2)/2]}\int_{-\infty}^{\infty}\frac{n[(1+w^2)+v\bar{x}w^2]}{n(1+w^2)+v\bar{x}w^2}[1+w^2]^{-(n-1)/2}\,dw$$

$$= \bar{x}^2 + \left(\frac{n-2}{n}\right)v\bar{x}^3I, \tag{5.10}$$

where

$$I = E\left[\frac{T^2}{(n-2)+T^2\bar{x}/\bar{x}_H}\right] \tag{5.11}$$

with $\bar{x}_H$ being the sample harmonic mean and the expectation is with respect to T which has the Student's t distribution with $(n-2)\,\text{df}$.

Thus the UMVUE of μ^3/λ is

$$n(n-1)\left(\frac{\bar{X}^3}{\hat{\lambda}}\right)I \tag{5.12}$$

where I is given in (5.11). The estimator in (5.12) is in a different form than given by Korwar. Although neither is in a closed mathematical form, the estimator in (5.12) is appealing because it is expressed in terms of expectation with respect to a known distribution, Student's t.

For $n = 2$, $I = \bar{X}_H/\bar{X}$ and the estimator is $2\bar{X}^2\bar{X}_H/\hat{\lambda}$. It can be easily seen that this is equal to the sample variance,

$$S^2 = \sum_{1}^{2}(X_i - \bar{X})^2.$$

Iwase and Seto (1983) obtained the UMVUE of the rth cumulant given in (2.8),

$$\hat{K}_r = \frac{\Gamma(r - 1/2)\Gamma((n - 1)/2)}{\sqrt{\pi}\,\Gamma((n-1)/2 + r + 1)}\, \bar{X}^{2r-1}V^{r-1}$$

$$\times F(r-1, r-\tfrac{1}{2}; [(n-1)/2]+r-1; -\bar{X}V/n), \qquad r \geq 1$$

$$(5.13)$$

where F stands for a Gauss hypergeometric function (Erdely, 1953). For $r = 2$, (5.13) again provides the UMVUE of the variance given by

$$\hat{K}_2 = [\bar{X}^3 V/(n - 1)]F(1, 3/2; (n + 1)/2, -\bar{X}V/n). \qquad (5.14)$$

Although the estimator in (5.14) is expressed in a closed form, this expression is similar to that in (5.12). Note that the part I/n in (5.12) is equivalent to F in (5.14).

Iwase and Seto also give the variance of the UMVUE in (5.14) and discuss that $\mathrm{Var}(\hat{K}_2)$ behaves as $(\mu^6/\lambda^2)(2 + 9\mu/\lambda)n^{-1} + 0(n^{-2})$ for large n. It is also pointed out that the MLE of the variance, $\hat{\mu}^3/\hat{\lambda}$, has the same asymptotic behavior.

Furthermore, these authors provide a result (their main theorem) that can be used to obtain the UMVUE of various parametric functions. Table 1 in their paper lists UMVUEs of several useful parametric functions of the inverse Gaussian distribution.

5.3   THE ROLE OF THE HARMONIC MEAN

The harmonic mean of sample observations plays an important role in statistical estimation of the inverse Gaussian parameters. This is not at all surprising because of the inverse relationship or the Brownian motion aspects discussed in Chapter 3. In an elementary situation the harmonic mean is sometimes used in averaging ratios, such as rates and prices. Many ratios involving variables x and y can be expressed either as x/y or y/x. The gasoline price can be stated as so many dollars per gallon or so many gallons per dollar. When the ratio is x/y, the harmonic mean is used in determining averages if x is held fixed and y is varying, whereas the arithmetic mean is considered when x is varying and y is fixed.

Suppose we wish to average k ratios $r_i = x_i/y_i$, $i = 1, 2, \ldots, k$, by $\sum x_i / \sum y_i$. First, assume the unit of y_i's is fixed, say y. Then $x_i = r_i y$ and the average $\sum x_i / \sum y_i = y \sum r_i / ky = \sum r_i / k = \bar{r}$, which is the arithmetic mean of the r_i. On the other hand, if the unit of x_i is fixed, say $x_i = x$ for all i, then $y_i = x/r_i$ and the average $\sum x_i / \sum y_i = kx/x \sum (1/r_i) = k / \sum (1/r_i) = \bar{r}_H$, which is the harmonic mean of the r_i.

The use of the harmonic mean is justified in computing the average drift v of a Brownian motion particle that has traveled a vertical distance s in time t by the equation $v = s/t$. This is because often a barrier is placed at a fixed distance and the time taken by a particle to reach the barrier is observed. Hence s is fixed and t is varying, resulting in a set of observations for time. Accordingly, we are averaging times per unit of distance rather than distance per unit of time. Thus the harmonic mean would be appropriate in computing the drift.

In a similar situation Fisher (1932) also pointed out that in computing the velocity of a fireball, the harmonic mean of a number of observations is more natural than the arithmetic mean to determine the average velocity. The reason for this is that the distance (path length) is computed by triangulation from the observed positions of beginning and ending, which are well defined, and hence is accurately measured. The time is largely a matter of retroactive estimation and is subject to error. Hence the distance unit can be regarded as fixed and the observed time as variable; thereby averaging times per unit distance (i.e., the harmonic mean) is justified.

Ferger (1931) discussed the use of the harmonic mean and gave an interesting account on this subject.

5.4 SAMPLING DISTRIBUTIONS

For a random sample of n observations $X_1, X_2, \ldots, X_n$ from $IG(\mu, \lambda)$, it was established in Theorem 4.7 that the sample mean $\bar{X} \sim IG(\mu, \lambda n)$, that the statistic $\sum (1/X_i - 1/\bar{X}) \sim (1/\lambda)\chi_{n-1}^2$, and that these statistics are independently distributed. Next we give some other statistical distributions analogous to the chi-square, t, and F distributions which arise in sampling from the normal.

Theorem 5.1 Let X_1, $X_2, \ldots, X_n$ be a random sample from $IG(\mu, \lambda)$ and let

$$\bar{X} = \frac{1}{n} \sum_1^n X_i, \quad V = \frac{1}{n-1} \sum_1^n \left(\frac{1}{X_i} - \frac{1}{X}\right).$$

Define

$$U = \frac{\sqrt{n}(\bar{X} - \mu)}{\mu[\bar{X}V]^{1/2}}. \tag{5.15}$$

Then U is distributed according to the density function given in (4.2) with λ replaced by $n\lambda$ and the degrees of freedom for the Student's t distribution is $(n-1)$ instead of n.

  Proof: The result follows easily from Theorem 4.4 because $\bar{X} \sim IG(\mu, n\lambda)$, $(n-1)V/n \sim (1/n\lambda)\chi^2_{n-1}$, and these statistics are independently distributed.

Corollary (i) The distribution of $|U|$ is the truncated (or folded) Student's t with $(n-1)$ df. (ii) $U^2 \sim F_{1,n-1}$, the Snedecor's F distribution with 1 and $(n-1)$ df.

  Proof: (i) and (ii) follow directly from the nature of the odd function in the expression of the density function of U.

  It may be mentioned that U is the uniformly most powerful unbiased (UMPU) test statistic for μ, as shown later in Section 6.2. However, its unconditional distribution discussed here in Theorem 5.1 involves λ/μ and is different from its conditional distribution, given $\sum X_i + \sum X_i^{-1}$, which is independent of this parameter. From the viewpoint of the analogy between the inverse Gaussian and the normal, this is a departure.

  Next we consider the two sample cases, where $X_1, X_2, \ldots, X_{n_1}$ and $Y_1, Y_2, \ldots, Y_{n_2}$ are two independent random samples from $IG(\mu, \lambda)$. Let

$$\bar{X} = \frac{1}{n_1} \sum_1^{n_1} X_i, \qquad V_1 = \sum_1^{n_1} \left(\frac{1}{X_i} - \frac{1}{\bar{X}} \right)$$

$$\bar{Y} = \frac{1}{n_2} \sum_1^{n_2} Y_i, \qquad V_2 = \sum_1^{n_2} \left(\frac{1}{Y_i} - \frac{1}{\bar{Y}} \right) \tag{5.16}$$

Then $\bar{X}$, $\bar{Y}$, V_1 and V_2 are jointly independent and $\bar{X} \sim IG(\mu, n_1\lambda)$, $\bar{Y} \sim I(\mu, n_2\lambda)$, $\lambda V_1 \sim \chi^2_{n_1-1}$ and $\lambda V_2 \sim \chi^2_{n_2-1}$. Clearly, the statistic

$$V_1 + V_2 \sim \left(\frac{1}{\lambda} \right) \chi^2_{n_1 + n_2 - 2}. \tag{5.17}$$

The following two results are obtained using these distributional properties.

Theorem 5.2   (a) The ratio $R = (n_2 - 1)V_1/(n_1 - 1)V_2 \sim F_{n_1-1, n_2-1}$, the F distribution with $(n_1 - 1)$ and $(n_2 - 1)$ df.

(b) $Q = \dfrac{V_1}{(V_1 + V_2)} \sim B\left(\dfrac{n_1 - 1}{2}, \dfrac{n_2 - 1}{2} \right)$,

the beta distribution with parameters $(n_1 - 1)/2$ and $(n_2 - 1)/2$.

Proof: The result in (a) follows because the variable R is the ratio of two independently distributed chi-square variates divided by their respective degrees of freedom. The variable Q in (b) can be written as

$$Q = \frac{mR}{1 + mR}, \qquad \text{where } m = \frac{n_1 - 1}{n_2 - 1}, \tag{5.18}$$

and is a one-to-one transformation, where Q increases from 0 to 1 as R increases from 0 to ∞. Thus the distribution function of Q is given by

$$P[Q < x] = P\left[R < \frac{m^{-1}x}{1 - x} \right] = F_{n_1-1, n_2-1}\left(\frac{m^{-1}x}{1 - x} \right).$$

Differentiating it with respect to x, one obtains the density function of Q given by

$$\frac{1}{B[(n_1 - 1)/2, (n_2 - 1)/2]} x^{(n_1 - 3)/2}(1 - x)^{(n_2 - 3)/2}, \qquad 0 < x < 1.$$

This establishes the result in (b).

Results (a) and (b) in Theorem 5.2 provide the distributions of test statistics later considered in Chapter 6 for testing the equality of two inverse Gaussian population scale parameters.

Next we consider the statistics which are useful in testing for the equality of two inverse Gaussian means.

Theorem 5.3   Let

$$\bar{Z} = \frac{n_1 \bar{X} + n_2 \bar{Y}}{n_1 + n_2}$$

and

$$Q_1 = n_1\left(\frac{1}{\bar{X}} - \frac{1}{\bar{Z}}\right) + n_2\left(\frac{1}{\bar{Y}} - \frac{1}{\bar{Z}}\right). \tag{5.19}$$

Then $\bar{Z} \sim I(\mu, (n_1 + n_2)\lambda)$ and $Q_1 \sim \chi_1^2/\lambda$, and $\bar{Z}$ and Q_1 are independently distributed.

Proof: Here $\bar{Z}$ is the combined average of the X_i and the Y_i, and Q_1 is the weighted sum of the deviations of $1/\bar{X}$ and $1/\bar{Y}$ each from $1/\bar{Z}$. Since $\bar{X}$ and $\bar{Y}$ are iid IG($\mu, n\lambda$), the result follows from an application of Theorem 4.7.

The two statistics, $V_1 + V_2$ and Q_1, are stochastically independent. Now due to (5.17), the statistic

$$\frac{(n_1 + n_2 - 2)Q_1}{V_1 + V_2} \sim F_{1, n_1 + n_2 - 2}. \tag{5.20}$$

As will be shown in the next chapter, $(\lambda Q_1)^{1/2}$ and $[(n_1 + n_2 - 2)Q_1/(V_1 + V_2)]^{1/2}$ are the UMPU test statistics for testing the equality of two inverse Gaussian means for the λ known and the λ unknown case, respectively. Their distributions parallel those of similar statistics in the one-sample case.

Tweedie (1957a) established a sampling distribution property for the inverse Gaussian similar to that for the normal distribution in a one-way classification. This result is given next.

Theorem 5.4 Let $X_{ij} \sim IG(\mu_i, \lambda), j = 1, 2, \ldots, n_i$ and $i = 1, 2, \ldots, k$, be independent rv's, and

$$\bar{X}_i = \frac{1}{n_i} \sum_1^{n_i} X_{ij}, \quad \bar{X} = \sum_{i=1}^k \sum_{j=1}^{n_i} \frac{X_{ij}}{n}, \quad n = \sum_1^k n_i,$$

$$R = \sum_{i=1}^k \sum_{j=1}^{n_i} \left(\frac{1}{X_{ij}} - \frac{1}{\bar{X}} \right), \quad R_1 = \sum_{i=1}^k \sum_{j=1}^{n_i} \left(\frac{1}{X_{ij}} - \frac{1}{\bar{X}_i} \right),$$

$$R_2 = \sum_{i=1}^k n_i \left(\frac{1}{\bar{X}_i} - \frac{1}{\bar{X}} \right). \tag{5.21}$$

Then $R = R_1 + R_2$. (5.22)

The two components R_1 and R_2 are stochastically independent with $R_1 \sim \lambda^{-1}\chi^2(n-k)$ and $R_2 \sim \lambda^{-1}\chi^2(k-1)$, provided the μ_i are equal. When $\mu_1 = \mu_2 = \cdots = \mu_k$, $R \sim \lambda^{-1}\chi^2(n-1)$.

Proof: The algebraic identity in (5.22) can be easily verified. Since the $X_{ij}, j = 1, 2, \ldots, n_i$, are iid $\sim IG(\mu_i, \lambda)$, the within-group statistic $\sum_{i=1}^{n_i} (1/X_{ij} - 1/\bar{X}_i) \sim \lambda^{-1}\chi^2(n_i-1)$ and is independent of $\bar{X}_i, i = 1, 2, \ldots, k$. Thus $R_1 \sim \lambda^{-1}\chi^2(n-k)$ and is stochastically independent of R_2.

When the μ_i are equal, say μ, then the $n_i\bar{X}_i \sim IG(n_i\mu, n_i\lambda)$, $i = 1, 2, \ldots, k$ and their combined average $\bar{X} = \sum_{i=1}^k n_i\bar{X}_i/n \sim IG(\mu, \lambda)$. Again, it follows that $R_2 \sim \lambda^{-1}\chi^2(k-1)$. Moreover it can be easily seen that $R \sim \lambda^{-1}\chi^2(n-1)$. This completes the proof.

The decomposition in terms of independent chi-square components as in (5.22) does not hold true in general, for the inverse Gaussian variates. For example, no analogy to the normal distribution theory exists in the case of two-way or higher levels of classification. However, in a special case of regression, Letac et al. (1985) established a decomposition of the chi-square deviance statistic into independent chi-squared components. This result given in the following theorem is analogous to that which exists for the normal distribution.

Theorem 5.5 Let $Y_i \sim \text{IG}(\mu_i, \lambda)$, $\mu_i = \beta/x_i$ with $x_i > 0$, $\beta > 0$, $i = 1, 2, \ldots, n$, be independent random variables. Define the $n - 1$ random variables

$$R_k = \lambda \left[\frac{1}{Y_i} + \frac{1}{T_{k-1}} - \frac{1}{T_k} \right], \qquad k = 2, \ldots, n \qquad (5.23)$$

where

$$T_j = \frac{\sum_{i=1}^{j} x_i^2 Y_i}{\sum_{i=1}^{j} x_i}, \qquad j = 1, 2, \ldots, n. \qquad (5.24)$$

Then (a) $R_2, \ldots, R_n$ and T_n are mutually independent, (b) $R_k \sim \chi_1^2$ for $k = 2, 3, \ldots, n$, and (c) $T_n \sim \text{IG}(\beta/\sum_1^n x_i, \lambda)$.

Proof: The result is proved by induction. Let

$$W_n = \exp\left(\sum_2^n s_i R_i \right).$$

For $n = 2$, it follows from a slight variant of the result in Theorem 4.7 that the conditional expectation,

$$E[W_2 \mid T_2] = (1 - 2s_2)^{-1/2}.$$

Assume that the following result holds true for $n - 1$ where $n > 2$.

$$E[W_{n-1} \mid T_{n-1}] = \prod_{i=2}^{n-1} (1 - 2s_i)^{-1/2}. \qquad (5.25)$$

Now we establish that the result in (5.25) holds true for n. Define $Z = \exp(s_n R_n)$. Then one can write

$$E[W_n \mid T_n] = E[W_{n-1} Z \mid T_n].$$

Since T_n is a function of T_{n-1} and Y_n because of

$$T_n = \alpha^2 Y_n + (1 - \alpha)^2 T_{n-1},$$

where $\alpha = x_n / \sum_1^n x_i$, and Z is a function of R_n (or, equivalently, of T_{n-1}, Y_n), it follows from the conditional expectation argument that

$$E[W_{n-1}Z \mid T_n] = E\{E(W_{n-1}Z \mid T_{n-1}, Y_n) \mid T_n\}$$

$$= E\{ZE(W_{n-1} \mid T_{n-1}) \mid T_n\}$$

$$= E(Z \mid T_n) \prod_{i=1}^{n-1} (1 - 2s_i)^{-1/2}$$

due to (5.25). Again, due to the result in Theorem 4.7, $E(Z \mid T_n) = (1 - 2s_n)^{-1/2}$. Accordingly, $E[W_n \mid T_n] = \prod_{i=1}^n (1 - 2s_i)^{-1/2}$. Hence R_2, $R_3, \ldots, R_n$ are mutually independent and each is distributed as chi-square with one degree of freedom. Also, these are independent of T_n, which is distributed as $IG(\beta / \sum x_i, \lambda)$.

It was once conjectured that an exact chi-squared decomposition, similar to Cochran's in the case of normal, is possible for the deviance statistics when a multiple regression model is assumed for the inverse Gaussian variables. This conjecture has been shown to be false by Letac and Seshadri (1986).

5.5 ESTIMATION FOR THE THREE-PARAMETER DISTRIBUTION

In Section 2.7 we introduced a three-parameter inverse Gaussian distribution. For a random sample $X_1, X_2, \ldots, X_n$ with each X_i distributed according to the density function in (2.22), the likelihood function is

$$L(\theta, \mu, \lambda) = \left(\frac{\lambda}{2\pi}\right)^{n/2} \prod_1^n (x_i - \theta)^{-3/2}$$

$$\times \exp\left[-\frac{\lambda}{2\mu^2} \sum_1^n \frac{[(x_i - \theta) - \mu]^2}{x_i - \theta}\right] I_{[x_{(1)} > \theta]}, \qquad (5.26)$$

where $I_{[\cdot]}$ is the indicator function and $x_{(1)}$ is the value of the smallest-order statistic. For a specified value of θ, the MLEs of the remaining two parameters μ and λ exist and are obtained from (5.2) by replacing X_i with $X_i - \theta$, $i = 1, 2, \ldots, n$, namely,

$$\hat{\mu}(\theta) = \bar{X} - \theta, \quad \hat{\lambda}(\theta) = \left[\frac{1}{n} \sum_{1}^{n} \left(\frac{1}{X_i - \theta} - \frac{1}{\bar{X} - \theta} \right) \right]^{-1}. \quad (5.27)$$

Thus, if the MLE of θ exists, it can be obtained by maximizing the partially maximized likelihood function $L^*(\theta) = L(\theta, \hat{\mu}(\theta), \hat{\lambda}(\theta))$ which is obtained from replacing μ and λ in (5.26) by $\hat{\mu}(\theta)$ and $\hat{\lambda}(\theta)$ given in (5.27). Suppose $\hat{\theta}$ is the MLE of θ. Then the MLEs of μ and λ are obtained from (5.27) by substituting $\hat{\theta}$ for θ.

It follows from (5.26) that

$$L^*(\theta) = \left[\frac{1}{n} \sum_{1}^{n} \frac{\bar{x} - x_i}{(x_i - \theta)(\bar{x} - \theta)} \right]^{-n/2} \prod_{1}^{n} (x_i - \theta)^{-3/2} I_{[x_{(1)} > \theta]}. \quad (5.28)$$

Differentiating $L^*(\theta)$ with respect to θ, we get

$$\frac{\partial L^*}{\partial \theta} = L^*(\theta) \left[\frac{3}{2} \sum_{1}^{n} (x_i - \theta) - \frac{n^{n/2+1}}{2} \right.$$

$$\left. \left(\frac{1}{\bar{x} - \theta} + \frac{\sum_{1}^{n} (\bar{x} - x_i)/(x_i - \theta)^2}{\sum_{1}^{n} (\bar{x} - x_i)/(x_i - \theta)} \right) \right].$$

Thus, $L^*(\theta)$ is maximized at $\hat{\theta}$, which is the solution of $(\partial L^*/\partial \theta) = 0$; or equivalently,

$$3(\bar{x} - \theta) = n^{n/2} \left[\frac{1}{\bar{x} - \theta} + \frac{\sum_{1}^{n} (\bar{x} - x_i)/(x_i - \theta)^2}{\sum_{1}^{n} (\bar{x} - x_i)/(x_i - \theta)} \right]. \quad (5.29)$$

Whether or not (5.29) has a solution for θ is not obvious. One may wonder about the condition under which the solution exists. Padgett and Wei (1979) gave a sufficient condition for the existence of MLE $\hat{\theta}$. They showed that if $n > 3$ and $\sum_{1}^{n} (x_i - \bar{x})^3 > 0$, then the MLE of θ exists and so do the MLEs of μ and λ. However, it is not a necessary condition as they themselves observed in their empirical studies. When $\sum_{1}^{n} (x_i - \bar{x})^3 < 0$, $L^*(\theta)$ can achieve its overall maximum at $\theta = -\infty$. Thus no MLE of the parameters is possible when the sample skewness is negative. In such a situation Cheng and Amin (1981) argued that as $\theta \to -\infty$, the three-parameter inverse Gaussian distribution approaches the normal, with MLE of the mean $\theta + \hat{\mu}(\theta)$ and the variance $\hat{\mu}^3(\theta)/\hat{\lambda}(\theta)$ to become the MLE of the mean and variance obtained from a normal sample. Because of this they suggested fitting the normal instead of the inverse Gaussian distribution when the sample skewness is less than $k(6/n)^{1/2}$, where k is chosen so that the probability of fitting the IG distribution when the true distribution is normal is small. However, as $n \to \infty$, $n^{-1} \sum (x_i - \bar{x})^3$ converges almost everywhere to μ_3, the third central moment of the IG distribution, where $\mu_3 > 0$, and thus MLEs exist for a large sample.

Next, Cheng and Amin (1981) proved that the MLEs $\hat{\theta}$, $\hat{\mu}$, and $\hat{\lambda}$ are consistent and possess the usual asymptotic normality property. They discussed the maximum likelihood (ML) estimation of IG, lognormal, and Weibull distributions for their three-parametric families of distributions and showed that (1) ML estimation broke down for the Weibull distribution when considered as a model for moderately to highly skewed data and (2) IG and lognormal distribution fits were almost identical except for highly skewed data, which fitted the IG model better than the lognormal. Because of these results and because the ML estimation can break down for the lognormal and the Weibull, they advocated the application of the inverse Gaussian distribution.

In the absence of MLEs of θ, μ, and λ, one may consider their estimation by the method of moments, where the first three moments of the distribution are equated to the corresponding sample moments and then solved for the parameters. In terms of cumulants of the distribution,

$$K_1 = \mu + \theta, \qquad K_2 = \frac{\mu^3}{\lambda}, \qquad K_3 = \frac{3\mu^5}{\lambda^2}$$

the estimates $\tilde{\theta}$, $\tilde{\mu}$, and $\tilde{\lambda}$ are given by

$$\tilde{\mu} + \tilde{\theta} = \bar{X}$$

$$\frac{\tilde{\mu}^3}{\tilde{\lambda}} = \frac{1}{n} \sum_{1}^{n} (X_i - \bar{X})^2, \qquad \frac{3\tilde{\mu}^5}{\tilde{\lambda}^2} = \frac{1}{n} \sum_{1}^{n} (X_i - \bar{X})^3.$$

Solving these equations for $\tilde{\theta}$, $\tilde{\mu}$, and $\tilde{\lambda}$, one gets

$$\tilde{\theta} = \bar{X} - \tilde{\mu}$$

$$\tilde{\mu} = \frac{3[(1/n)\sum_1^n (X_i - \bar{X})^2]^2}{(1/n)\sum_1^n (X_i - \bar{X})^3}$$

$$\tilde{\lambda} = \frac{\tilde{\mu}^3}{(1/n)\sum (X_i - \bar{X})^2}. \tag{5.30}$$

It can be shown that these estimators, $\tilde{\theta}$, $\tilde{\mu}$, $\tilde{\lambda}$, are consistent and asymptotically normally distributed (Padget and Wei, 1979). It follows that as $n \to \infty$, the distribution of $n^{1/2}(\tilde{\theta} - \theta, \tilde{\mu} - \mu, \tilde{\lambda} - \lambda)$ tends to trivariate normal with mean vector $\mathbf{0}$ and dispersion matrix

$$\Sigma = \mu^2 \begin{bmatrix} p & -p & -q \\ -p & p + 1/\phi & q \\ -q & q & r \end{bmatrix}, \tag{5.31}$$

where

$$p = \frac{2(\phi + 6)^2}{3\phi}$$

$$q = 2(\phi + 5)(\phi + 8)$$

$$r = 2\phi[\phi + 3(\phi + 5)(\phi + 9)]. \tag{5.32}$$

The expressions in (5.31) and (5.32) are due to Jones and Cheng (1984). When the MLE exists, they also showed that the corresponding dispersion matrix of the ML estimators is

$$\Sigma^* = A \begin{bmatrix} 1 & -1 & -B \\ -1 & C & B \\ -B & B & D \end{bmatrix}, \tag{5.33}$$

where

$$A = \frac{2\phi^2\mu^2}{3(\phi + 4)} \qquad C = \frac{2\phi^3 + 3\phi + 12}{2\phi^3}$$

$$B = 3(\phi + 1) \qquad D = 3(3\phi^2 + 7\phi + 7). \tag{5.34}$$

Thus the joint asymptotic efficiency of the moment estimators $(\tilde{\theta}, \tilde{\mu}, \tilde{\lambda})$ defined by the ratio of two determinants, $|\Sigma^*|/|\Sigma|$ is equal to $\phi^4/(\phi + 4)[\phi(\phi + 6)^2 + 12(\phi + 5)]$. Similarly, the asymptotic relative efficiencies of the moment estimators of individual parameters can be obtained and shown to depend only on the shape parameter ϕ. Jones and Cheng (1984) have computed the relative efficiencies for a range of values of ϕ. Their numerical results show that when the inverse Gaussian distribution is fairly skewed, that is, ϕ is small, the relative efficiencies are low; but as $\phi \to \infty$, the relative efficiency in each case approaches 1. The latter corresponds to the situation where the inverse Gaussian distribution tends to the normal distribution for which the moment and the ML estimators are the same.

5.6 GOODNESS OF FIT

The Kolmogorov-Smirnov statistics can be used to evaluate the goodness of fit of the inverse Gaussian model prior to its application for data analysis. The critical values of these statistics are known provided a hypothesized distribution is completely specified. Otherwise, the percentage points are distribution dependent. Here we give certain percentage points of the Kolmogorov-Smirnov statistics applicable to the inverse Gaussian distribution.

Given a random sample of size n, the Kolmogorov-Smirnov statistic is defined by

$$D = \max(D^+, D^-) \tag{5.35}$$

where

$$D^+ = \max_{1 \leq i \leq n} [i/n - F(x)] \quad D^- = \max_{1 \leq i \leq n} [F(x) - (i - 1)/n]$$

Here $F(x)$ denotes the hypothesized distribution function. We considered the following form for $F(x)$ in our computations of the cdf:

$$F(z) = \Phi(z/[1 + z\phi^{-1/2}]^{1/2})$$
$$+ e^{2\phi}\Phi((z + 2\phi^{1/2})/[1 + z\phi^{-1/2}]^{1/2}) \tag{5.36}$$

where

$$z = \phi^{1/2}(x/\mu - 1) = \lambda/\mu$$

Considering $\alpha = .20, .15, .10, .05, .025$ and $.01$, the $(1 - \alpha)$ percentage points were estimated for statistic D using Monte Carlo simulations. The inverse Gaussian distributions were considered corresponding to $\phi = .25, .5, 1, 2, 4, 8$ and 16; this covered a wide range from highly skewed to nearly normal type densities. Ten thousand samples were generated for these distributions using the procedure described in Sec. 4.5. The percentage points of D were estimated in each case for sample size $n = 5, 6, \ldots, 35$, as well as $n = 40 (10) 100$. Table 5.1 contains the estimated $.95$ percentage points of D corresponding to various combinations of ϕ and n.

The use of results such as given in Table 5.1 is limited to certain selected values of ϕ and n. To cover all possible sample sizes and any calculated ϕ value, the estimated D values were smoothed as functions of n and ϕ. Plots of D versus $1/\sqrt{n}$ and $1/\sqrt{\phi}$ were obtained. These plots indicated a distinction between functions for $\phi \leqslant 1$ and $\phi > 1$. Because of this difference, the data were smoothed where two sets of equations corresponding to $\phi \leqslant 1$ and $\phi > 1$ were obtained for each α level. These equations are of the form,

$$D = \beta_0 + \beta_1 \sqrt{n} + \beta_2 \sqrt{\phi} \tag{5.37}$$

Table 5.1 .95 Percentile of Kolmogorov-Smirnov Statistic

n \ ϕ	.25	.50	1	2	4	8	16
5	0.5224	0.4318	0.3855	0.3655	0.3604	0.3599	0.3598
6	0.4908	0.4186	0.3603	0.3358	0.3278	0.3276	0.3280
7	0.4618	0.3914	0.3394	0.3145	0.3085	0.3091	0.3101
8	0.4433	0.3669	0.3227	0.3007	0.2927	0.2914	0.2927
9	0.4294	0.3568	0.3090	0.2851	0.2779	0.2764	0.2772
10	0.4045	0.3428	0.2928	0.2714	0.2636	0.2634	0.2629
11	0.3991	0.3322	0.2854	0.2687	0.2536	0.2516	0.2511
12	0.3662	0.3202	0.2764	0.2524	0.2430	0.2413	0.2417
13	0.3748	0.3088	0.2672	0.2446	0.2348	0.2329	0.2338
14	0.3652	0.3039	0.2602	0.2380	0.2286	0.2280	0.2283
15	0.3599	0.2983	0.2544	0.2322	0.2238	0.2202	0.2196
16	0.3429	0.2884	0.2468	0.2259	0.2168	0.2146	0.2141
17	0.3405	0.2785	0.2400	0.2180	0.2077	0.2071	0.2071
18	0.3325	0.2986	0.2365	0.2142	0.2039	0.2021	0.2016
19	0.3266	0.2715	0.2303	0.2088	0.1996	0.1971	0.1967
20	0.3175	0.2653	0.2266	0.2040	0.1939	0.1904	0.1911
25	0.2960	0.2464	0.2080	0.1867	0.1765	0.1743	0.1739
50	0.2292	0.1897	0.1583	0.1381	0.1281	0.1254	0.1247
100	0.1812	0.1473	0.1199	0.1014	0.0926	0.0894	0.0888

Source: Stark and Chhikara, 1988.

Table 5.2   Coefficients by α level and ϕ class

	$\phi \leq 1$			$\phi > 1$		
α	β_0	β_1	β_2	β_0	β_1	β_2
.200	-0.048	0.7381	0.0668	0.004	0.6631	0.0206
.150	-0.052	0.7736	0.0727	0.005	0.6909	0.0218
.100	-0.057	0.8173	0.0791	0.006	0.7254	0.0237
.050	-0.064	0.8864	0.0907	0.008	0.7773	0.0269
.025	-0.069	0.9498	0.0998	0.011	0.8215	0.0297
.010	-0.070	1.0170	0.1086	0.014	0.8689	0.0355

Source: Stark and Chhikara, 1988.

with the coefficient β_0, β_1, and β_2 shown in Table 5.2 for each class of ϕ values. If ϕ is large, say $\phi > 100$, the equations from Stephens (1974) for the normal distribution should be used since for ϕ large the inverse Gaussian converges to the normal.

The accuracy of equation (5.37) can be demonstrated using the values in Table 5.1. For example, assuming a sample size of $n = 16$, $\phi = 1$, and $\alpha = .05$, the equation yields $D = .2438$. The corresponding value in Table 1 is .2468. The absolute difference of .0015 contributes to an overall mean-squared error value of 6.8×10^{-5}. For the power of the Kolmogorov-Smirnov test and other details, refer to Stark and Chhikara (1988).

5.7   AN EXAMPLE

Lieblein and Zelen (1956) analyzed certain test data on the endurance of deep groove ball bearings. The data consist of the number of million revolutions before failure for each of 23 ball bearings used in the life test and are given here.

17.88	28.92	33.00	41.52	42.12	45.60	48.48
51.84	51.96	54.12	55.56	67.80	68.64	68.64
68.88	84.12	93.12	98.64	105.12	105.84	127.92
128.04	173.40					

These data have been previously analyzed assuming they follow a Weibull or a lognormal distribution. A probability plot of the data (Figure 5.1) shows them also to be consonant with an inverse Gaussian model. We will use them to illustrate the parametric

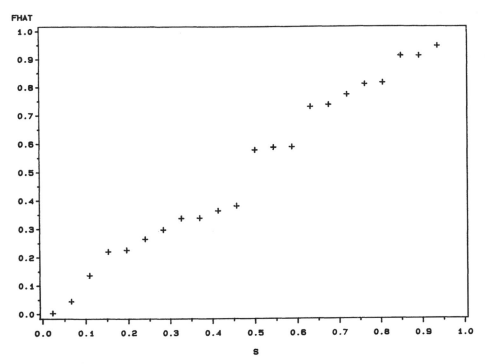

FIGURE 5.1 Plot of MLE and empirical CDF for the ball bearing failure data.

estimation and some of the hypothesis-testing methods discussed in the monograph.

The MLEs of μ, λ, and $\sigma^2 = \mu^3/\lambda$ are

$$\hat{\mu} = 72.22, \qquad \hat{\lambda} = 231.63, \qquad \hat{\sigma}^2 = 1626.15$$

The MLE of the cdf, $\hat{F}(x)$, was obtained by replacing the parameters μ and λ in (2.14) by their MLEs. The empirical cdf, $S(x) = (i - \frac{1}{2})/n$, was obtained corresponding to each observed x value. It was found that the observed value of the Kolmogorov–Smirnov statistic is 0.0994, indicating a good fit of data to the inverse Gaussian model.

6

Significance Tests

6.1 INTRODUCTION

In Chapter 5 we introduced several statistics and pointed out their potential uses for testing hypotheses concerning the parameters μ and λ. Their exact sampling distribution properties make them the main candidates for obtaining exact significance tests for the parameters. In this chapter we derive certain optimum tests in their exact forms and show that these have a close analogy with their counterparts for the normal distribution; tests for the scale parameter λ are based on the percentage points of chi-square distributions, and tests for the mean μ are determined by the percentage points of standard normal and Student's t distributions. The analogy is shown to persist in the case of two-sample tests.

6.2 ONE-SAMPLE METHODS

6.2.1 Introduction

For a given random sample of size n, $\mathbf{X} = (X_1, X_2, \ldots, X_n)$, from IG($\mu, \lambda$), the joint density function of $\mathbf{X}$ is

$$f(\mathbf{x}; \mu, \lambda) = \left(\frac{\lambda}{2\pi}\right)^{n/2} e^{n\lambda/\mu} \prod_1^n x_i^{-3/2}$$

$$\times \exp\left(-\frac{\lambda}{2\mu^2}\sum x_i - \frac{\lambda}{2}\sum \frac{1}{x_i}\right). \tag{6.1}$$

When either μ or λ is specified, the density function in (6.1) reduces to a one-parameter exponential family and therefore, possesses the monotone likelihood ratio (MLR) property. So if we consider one-sided hypotheses for each parameter with the other parameter specified, we obtain uniformly most powerful (UMP) tests. In the case of a hypothesis with a two-sided alternative, the tests are UMP-unbiased. All these optimum tests are discussed next.

6.2.2   Tests for μ

Suppose one considers testing the following hypotheses:

$$H_0^{(1)}: \mu \leqslant \mu_0 \text{ against } H_A^{(1)}: \mu > \mu_0$$

and

$$H_0^{(2)}: \mu = \mu_0 \text{ against } H_A^{(2)}: \mu \neq \mu_0. \tag{6.2}$$

λ Known

It follows from the MLR property of (6.1) that a UMP level α test of $H_0^{(1)}$ against $H_A^{(1)}$ exists and has the critical region given by $\bar{X} > k$ where k is determined according to

$$\int_k^\infty g(t)\, dt = \alpha, \tag{6.3}$$

where g denotes the pdf of $\bar{X} \sim \text{IG}(\mu_0, n\lambda)$. However, no UMP test of $H_0^{(2)}$ versus $H_A^{(2)}$ exists; but one can obtain a UMP-unbiased test instead [see Lehmann (1959) Chap. IV]. The critical region of a UMP-unbiased level α test is $\bar{X} < k_1$, or $\bar{X} > k_2$, where k_1 and k_2

are the solutions of

$$\int_{k_1}^{k_2} g(t)\,dt = 1 - \alpha \qquad \text{and} \qquad \int_{k_1}^{k_2} tg(t)\,dt = (1 - \alpha)\mu_0. \qquad (6.4)$$

We consider the following transformation of $\bar{X}$ as it facilitates the solution of the equation in (6.3) for k and of the equations in (6.4) for k_1 and k_2:

$$Y = \frac{\sqrt{n\lambda}(\bar{X} - \mu_0)}{\mu_0\sqrt{\bar{X}}}. \qquad (6.5)$$

Since $\bar{X} \sim IG(\mu_0, n\lambda)$, it follows from Theorem 2.1 in Chapter 2 that the UMP critical region, $\bar{X} > k$, corresponds to

$$Y > C \qquad (6.6)$$

where C is given by

$$\Phi(C) + e^{2n\lambda/\mu_0}\Phi\sqrt{\frac{4n\lambda}{\mu_0} + C^2} = 1 - \alpha, \qquad (6.7)$$

Φ being the standard normal distribution function. Thus, for a given α, one can obtain C from (6.7) by iteration using probabilities from a standard normal table.

Next the UMP-unbiased critical region, $\bar{X} < k_1$ or $\bar{X} > k_2$, corresponds to $Y < C_1$ or $Y > C_2$ where, due to (6.4), C_1 and C_2 are given by

$$\int_{C_1}^{C_2} \left[1 - \frac{y}{\sqrt{4n\lambda/\mu_0 + y^2}}\right]\left(\frac{1}{\sqrt{2\pi}}\right)\exp\left(\frac{-y^2}{2}\right)dy = 1 - \alpha$$

and

$$\int_{C_1}^{C_2} \left[1 + \frac{y}{\sqrt{4n\lambda/\mu_0 + y^2}}\right]\left(\frac{1}{\sqrt{2\pi}}\right)\exp\left(\frac{-y^2}{2}\right)dy = 1 - \alpha;$$

the constants C_1 and C_2 are nonzero unless $\alpha = 1$. Consequently, we obtain $C_1 = -C_2 = z_{1-\alpha/2}$, the $100(1 - \alpha/2)$ percentage point of the standard normal distribution. Hence the critical region of a UMP-unbiased level α test of $H_0^{(2)}: \mu = \mu_0$ against $H_A^{(2)}: \mu \neq \mu_0$, λ known, corresponds to

$$|Y| > z_{1-\alpha/2}, \tag{6.8}$$

where Y is given in (6.5).

λ Unknown

Employing a more natural parameterization, the exponential family in (6.1) can be written in the form:

$$\prod_1^n f(x_i; \mu, \lambda) = C(\theta, \psi)\left(\prod_1^n x_i^{-3/2}\right)$$

$$\times \exp\left[\theta \sum_1^n x_i + \psi \sum_1^n (x_i + x_i^{-1})\right],$$

where $\theta = \lambda(1 - \mu^{-2})/2$, $\psi = -\lambda/2$. Since $f(x; \mu, \lambda) = \mu^{-1}f(x/\mu; 1, \lambda/\mu)$ for the inverse Gaussian density function, without loss of generality, assume $\mu_0 = 1$ in (6.2). Then these hypotheses can equivalently be stated as follows:

$$H_0'^{(1)}: \theta \leqslant 0 \text{ against } H_A'^{(1)}: \theta > 0$$

and

$$H_0'^{(2)}: \theta = 0 \text{ against } H_A'^{(2)}: \theta \neq 0.$$

For a given level α, UMP-unbiased tests of these hypotheses exist (Lehmann, 1959); the critical region of a UMP-unbiased level α test of $H_0'^{(1)}$ versus $H_A'^{(1)}$ is $U > k$, where $U = \sum_1^n X_i$ and k is determined by

$$\int_k^\infty h(u \mid t) \, du = \alpha \tag{6.9}$$

for all $t = \sum (x_i + x_i^{-1})$. In (6.9) $h(u|t)$ denotes the conditional density function of U given t when $\theta = 0$ and is given by

$$h(u|t) = \frac{n}{B[\frac{1}{2}, (n-1)/2]} \frac{1}{\sqrt{u^3(t-2n)}}$$

$$\times \left[1 - \frac{(u-n)^2}{u(t-2n)} \right]^{(n-3)/2}, \qquad 0 < \frac{(u-n)^2}{u(t-2n)} < 1.$$

$$(6.10)$$

In the case of testing $H_0^{(2)}$ versus $H_A^{(2)}$, the UMP-unbiased critical region corresponds to $U < k_1$, or $U > k_2$, where k_1 and k_2 are determined by

$$\int_{k_1}^{k_2} h(u|t)\, du = 1 - \alpha$$

and

$$\int_{k_1}^{k_2} uh(u|t)\, du = (1 - \alpha) \int_{-\infty}^{\infty} uh(u|t)\, du. \qquad (6.11)$$

Let

$$W = \frac{\sqrt{(n-1)}(U - n)}{\left\{ U(T - 2n) \left[1 - \frac{(U-n)^2}{U(T-2n)} \right] \right\}^{1/2}}.$$

It is a one-to-one transformation and the conditional density function of W given $T = t$ is (Chhikara and Folks, 1976)

$$p(w|t) = \frac{1}{\sqrt{n-1}\, B[\frac{1}{2}, (n-1)/2]} \left[1 - \frac{w\sqrt{(t-2n)(n-1)}}{\sqrt{4n + (t+2n)w^2/(n-1)}} \right]$$

$$\times \left[1 + \frac{w^2}{n-1} \right]^{-n/2}, \qquad -\infty < w < \infty.$$

When expressed in terms of sample observations,

$$W = \frac{\sqrt{(n-1)}(\bar{X}-1)}{\sqrt{\bar{X}V}} \tag{6.12}$$

where $V = \sum_1^n (1/X_i - 1/\bar{X})/n$.

After making substitutions in (6.11) it follows that the critical region, $U > k$ corresponds to $W > C$ where C is given by

$$F_{t,n-1}(-C) + \left[\frac{t+2n}{t-2n}\right]^{(n-2)/2} F_{t,n-1}(-\sqrt{4n+(t+2n)C^2}) = \alpha \tag{6.13}$$

where $F_{t,n-1}$ is the Student's t distribution function with $(n-1)$ degrees of freedom and $t = \sum_1^n (x_i + x_i^{-1})$. Now for a specified α and known t, one can easily find C from Student's t tables by iteration because both $F_{t,n-1}(-C)$ and $F_{t,n-1}[-\sqrt{4n+(t+2n)C^2}]$ are monotonically nonincreasing with respect to C. Moreover, (6.13) can be written in terms of the beta distribution, for which extensive tables are available, and hence iteration can be more accurately carried out.

In the general case for testing $H_0^{(1)}: \mu \leqslant \mu_0$ against $H_A^{(1)}: \mu > \mu_0$, λ unknown, a UMP-unbiased level α test is obtained by replacing X_i by X_i/μ_0, $i = 1, 2, \ldots, n$, in (6.12) with the test statistics given by

$$\frac{\sqrt{(n-1)}(\bar{X} - \mu_0)}{\mu_0\sqrt{\bar{X}V}} \tag{6.14}$$

and the critical point C of the significance test is the solution of the following equation:

$$F_{t,n-1}(-C) + \left[\frac{\sum_1^n (x_i + \mu_0)^2/x_i}{\sum_1^n (x_i - \mu_0)^2/x_i}\right]^{(n-2)/2}$$

$$\times F_{t,n-1}\left[-\sqrt{4n + C^2\mu_0 \sum_1^n \frac{(x_i + \mu_0)^2}{x_i}}\right] = \alpha. \tag{6.15}$$

In the two-sided case, the transformation from U to W simplifies (6.11) resulting in a t test. Chhikara and Folks (1976) showed that the critical region of a UMP-unbiased level α test of $H_0^{(2)}:\theta = 0$ against $H_A^{(2)}:\theta \neq 0$ is $|W| > t_{1-\alpha/2}$ or, in the general case of testing $H_0^{(2)}:\mu = \mu_0$ against $H_A^{(2)}:\mu \neq \mu_0$, λ unknown, this critical region is

$$\left| \frac{\sqrt{(n-1)}(\bar{X} - \mu_0)}{\mu_0\sqrt{\bar{X}V}} \right| > t_{1-\alpha/2} \tag{6.16}$$

where $t_{1-\alpha/2}$ is the $100(1 - \alpha/2)$ percentage point of the Student's t distribution with $(n - 1)$ degrees of freedom.

It is interesting to see that a two-sided alternative test is simply the normal or t test, depending on whether λ is known or unknown. Though such simplicity is lacking in the case of a one-sided alternative, the test can still be easily performed with access to tables of percentage points of the standard normal or Student's t distributions. Also, when n is large and/or if λ compared with μ is large, the critical value in (6.7) (or (6.13)) can be well approximated by $z_{1-\alpha}$ (or $t_{1-\alpha}$). Furthermore, Davis (1980) showed that the optimum tests for the two-sided alternative case are equivalent to their corresponding likelihood ratio tests. In the one-sided case the later ones coincide with our approximate optimum tests based on $z_{1-\alpha}$ and $t_{1-\alpha}$.

Optimum tests in the special case of testing $H_0:\mu = \infty$ versus $H_A:\mu < \infty$ were given by Nadas (1973) assuming λ known and by Seshadri and Shuster (1974) for the λ unknown case. Those tests are special cases of the optimum tests discussed here. For a further discussion one should refer to Chhikara and Folks (1976).

The power functions of these tests have not yet been obtained. In the special case of $H_0:\mu = \infty$, Patil and Kovner (1976) derived the power functions of these tests in an exact form. Looking at the complexity of their result, our feeling is that though the powers of these tests can be obtained using the related distributions given earlier in Chapter 4 and in Chhikara and Folks (1975), the exact power functions will be too complex to provide simple numerical computations and to add any significance to the statistical methodology presented here.

6.2.3   Tests for λ

When μ *is known*, the statistic $T = \sum_1^n (X_i - \mu)^2/X_i$ is sufficient for λ and $(\mu^2/\lambda)T \sim \chi_n^2$. It follows that an UMP-unbiased test of the hypothesis $H_0:1/\lambda = 1/\lambda_0$ versus $H_A:1/\lambda \neq 1/\lambda_0$ has the critical region given by $T \leqslant C_1$ or $T \geqslant C_2$ where, for a given level α of the test, C_1 and C_2 are determined by

$$\int_{C_1}^{C_2} g_n(t)\, dt = (1 - \alpha) \quad \text{and} \quad \int_{C_1}^{C_2} t g_n(t)\, dt = n(1 - \alpha) \quad (6.17)$$

with $g_n(t)$ denoting the density function of χ_n^2. It is easy to show that $t g_n(t) = n g_{n+2}(t)$. Thus the two conditions in (6.17) can be written as

$$F_{\chi_n^2}(C_2) - F_{\chi_n^2}(C_1) = F_{\chi_{n+2}^2}(C_2) - F_{\chi_{n+2}^2}(C_1) = 1 - \alpha \quad (6.18)$$

where $F_{\chi_n^2}$ denotes the chi-square distribution function with n degrees of freedom. Though C_1 and C_2 are uniquely determined from (6.18) using tables of the chi-square distribution, an iterative method is required. On the other hand, when n is large or λ compared with μ is large, the equal tail test given by

$$F_{\chi_n^2}(C_1) = 1 - F_{\chi_n^2}(C_2) = \frac{\alpha}{2} \quad (6.19)$$

provides a good approximation of the exact test.

μ Unknown

Roy and Wasan (1968) derived the UMP-unbiased test for $H_0:1/\lambda = 1/\lambda_0$ versus $H_A:1/\lambda \neq 1/\lambda_0$. The approach is the same as described above for the case of μ known. The statistic $\sum_1^n (1/X_i - 1/\bar{X})$ is distributed as $(1/\lambda)\chi_{n-1}^2$. The exact or approximate test follows from (6.18) or (6.19) by considering $\sum_1^n (1/X_i - 1/\bar{X})$ in place of T and replacing n by $(n-1)$.

The optimum tests for one-sided hypotheses are straightforward to obtain and are parallel to those for the variance of the normal distribution as seen above in the case of hypotheses with two-sided alternatives. The two-sided tests are also the likelihood ratio tests (Davis, 1980).

6.2.4 Confidence Intervals

From the optimum tests obtained in Section 6.2.2 one can obtain uniformly most accurate (UMA) or UMA-unbiased confidence intervals for the parameters by inverting their acceptance regions. This follows directly due to the one-to-one relationship between the acceptance region of a test and the corresponding confidence intervals for the parameter involved in the hypothesis testing (Lehmann, 1956).

When λ *is known*, it follows from (6.8) that its $100(1 - \alpha)$ percent confidence interval for μ is

$$\bar{X}\left[1 + \sqrt{\frac{\bar{X}}{n\lambda}}z_{1-\alpha/2}\right]^{-1}, \bar{X}\left[1 - \sqrt{\frac{\bar{X}}{n\lambda}}z_{1-\alpha/2}\right]^{-1} \tag{6.20}$$

provided the quantity $1 - \sqrt{\bar{X}/n\lambda}\,z_{1-\alpha/2}$ is positive. Otherwise one gets the one-sided interval,

$$\left(\bar{X}\left[1 + \sqrt{\frac{\bar{X}}{n\lambda}}z_{1-\alpha/2}\right]^{-1}, \infty\right)$$

In the case of λ *unknown*, such a confidence interval for μ follows from (6.16) and is given by

$$\left(\bar{X}\left[1 + \sqrt{\frac{\bar{X}V}{(n-1)}}t_{1-\alpha/2}\right]^{-1}, \bar{X}\left[1 - \sqrt{\frac{\bar{X}V}{(n-1)}}t_{1-\alpha/2}\right]^{-1}\right)$$

$$\tag{6.21}$$

if

$$1 - \sqrt{\frac{\bar{X}V}{(n-1)}}t_{1-\alpha/2} > 0 \quad \text{and} \quad \left(\bar{X}\left[1 + \sqrt{\frac{\bar{X}V}{(n-1)}}t_{1-\alpha/2}\right]^{-1}, \infty\right)$$

otherwise.

A lower confidence bound for μ can be obtained from the one-sided hypothesis test discussed in Section 6.2.2. However, it is not possible to give any explicit bounds because determination of the critical points involves iteration. Instead, good approximate bounds can be easily obtained by observing that whenever n or λ compared with μ is large, the second terms on the left in (6.7) and (6.13) are negligible. For the $100(1 - \alpha)$ percent confidence level, these approximate lower bounds are given by

$$\bar{X}\left[1 + \sqrt{\frac{\bar{X}}{n\lambda}}\, z_{1-\alpha}\right]^{-1} \tag{6.22}$$

for the λ *known* case and

$$\bar{X}\left[1 + \sqrt{\frac{\bar{X}V}{(n-1)}}\, t_{1-\alpha}\right]^{-1} \tag{6.23}$$

for the λ *unknown* case.

The construction of confidence intervals for λ involves simple inversion of the chi-square probability statements paralleling that for the variance of the normal distribution.

6.2.5 Numerical Examples

Example 1

Consider the failure times of 23 ball bearings given in Section 5.7 and test $H_0 : \mu = 70$ against $H_A : \mu \neq 70$, λ unknown, for the mean failure time. From the data, $\bar{x} = 72.22$ and $v = .00432$. It is seen that the observed value of the test statistic given in (6.14) is .272 and is much smaller than the 5 percent significance level critical value, $t_{22,.975} = 2.074$; thus H_0 is not rejected.

On the other hand, suppose the null hypothesis is $H_0 : \mu = 57$ against the alternative hypothesis $H_A : \mu \neq 57$. In this case the observed test statistic is 2.29, which exceeds the above critical value and hence H_0 is rejected.

Or one might consider testing $H_0 : \mu \leqslant 70$ against $H_A : \mu > 70$. In this case the critical value C is obtained by solving (6.15) with $\alpha = .05$. After substitution in (6.15), one finds that the second term is almost zero and can be neglected. Accordingly, taking

$F_{t,22}(C) = .95$, it yields $C = 1.717$. Since the observed value, .272, is less than the critical value, the one-sided hypothesis is not rejected.

Example 2

We consider the failure data for 10 identical devices tested under high stress conditions and used by Nadas (1973) for testing what he called zero drift in Brownian motion. On the basis of this data, the hypothesis of zero drift, that is, infinite mean life for the device, was rejected. Suppose we consider testing $H_0: \mu = 1$ against $H_A: \mu \neq 1$, λ unknown, using the same data for which $\bar{x} = 1.352$, $\Sigma(1/x_i - 1/\bar{x}) = 2.083$ and $t = \sum x_i + \sum x_i^{-1} = 22.999$. Now making substitution in (6.12), the observed value of the test statistic W is 1.99. Since $t_{9,.975} = 2.262$, we do not reject the hypothesis because the observed value falls outside the critical region. On the other hand, let us test $H_0: \mu \leq 1$ against $H_A: \mu > 1$. Now compare the observed value of 1.99 with the critical value C given by (6.13) or, after substitution, by

$$F_{t,9}(-C) + \left(\frac{43}{3}\right)^4 F_{t,9}(-\sqrt{40 + 43C^2}) = .05. \qquad (6.24)$$

Since $F_{t,9}(-1.833) = .05$, the left side is greater than .05 for $C = 1.833$. Accordingly, $C > 1.833$. It is easy to find that for $C = 1.90$, the left side in (6.24) is equal to .05 up to the third decimal place, and therefore the critical point can be taken equal to 1.90. Since it places the observed value of 1.99 in the critical region, the hypothesis $H_0: \mu \leq 1$ is rejected.

One may need to use the extended Student's t tables [e.g., see Federighi (1959), who has tabulated high-level percentage points for the Student's t] in order to maintain accuracy in computation of the critical value C.

Next the 95 percent confidence interval for μ is (.966, 2.254). Using the critical value of 1.90 obtained above, the exact 95 percent confidence lower bound for μ is 1.012. On the other hand, from (6.23), the approximate 95 percent confidence lower bound is 1.021, which is close to the exact.

6.3 TWO-SAMPLE METHODS

6.3.1 Test for the Equality of Two Inverse Gaussian Means

Chhikara (1975) derived UMP-unbiased tests for the equality of two inverse Gaussian population means, say μ and v, assuming common scale parameter λ. Both one-sided and two-sided alternative hypotheses were discussed by the author. The two-sample results given in Chapter 5 suggest that optimum tests can be constructed and should parallel those in the one-sample case; this, in fact, is true. Derivation of the UMP-unbiased tests for the equality of μ and v is fairly straightforward and hence is omitted. Here we simply give the critical region of each UMP-unbiased test.

Consider the testing of the following hypotheses:

$$H_0^1 : \mu \leqslant v \qquad \text{versus} \qquad H_A^1 : \mu > v$$

and

$$H_0^2 : \mu = v \qquad \text{versus} \qquad H_A^2 : \mu \neq v. \tag{6.24}$$

When λ is *known*, the rejection region of a UMP-unbiased level α test of H_0^1 versus H_A^1 is

$$\frac{(n_1 n_2 \lambda)^{1/2}(\bar{X} - \bar{Y})}{[\bar{X}\bar{Y}(n_1\bar{X} + n_2\bar{Y})]^{1/2}} > C, \tag{6.26}$$

where C is the solution of the equation

$$\Phi(C) + \frac{n_2 - n_1}{n_1 + n_2} \exp\left(\frac{2n_1 n_2 \lambda}{n_1\bar{X} + n_2\bar{Y}}\right)[1 - \Phi(C')] = 1 - \alpha \tag{6.27}$$

with

$$C' = \left[C^2 + \frac{4n_1 n_2 \lambda}{n_1\bar{X} + n_2\bar{Y}}\right]^{1/2}.$$

In the case of the two-sided alternative, that is, H_0^2 versus H_A^2, λ known, a UMP unbiased level α test reduces to a normal test and its rejection region is given by

$$\left| \frac{(n_1 n_2 \lambda)^{1/2}(\bar{X} - \bar{Y})}{[\bar{X}\bar{Y}(n_1\bar{X} + n_2\bar{Y})]^{1/2}} \right| > z_{1-\alpha/2}. \tag{6.28}$$

When λ is *unknown*, the rejection region of a UMP-unbiased level α test of H_0^1 versus H_A^1 is

$$\frac{[n_1 n_2(n_1 + n_2 - 2)]^{1/2}(\bar{X} - \bar{Y})}{[\bar{X}\bar{Y}(n_1\bar{X} + n_2\bar{Y})(V_1 + V_2)]^{1/2}} > C, \tag{6.29}$$

where $V_1 = \sum_1^{n_1}(1/X_i - 1/\bar{X})$, $V_2 = \sum_1^{n_2}(1/Y_i - 1/\bar{Y})$, and C is the solution of the following equation:

$$F_{t,n-2}(C) - \frac{n_1 - n_2}{n_1 + n_2}\rho[1 - F_{t,n-2}(C')] = 1 - \alpha \tag{6.30}$$

with

$$C' = \left[C^2 + \frac{4n_1 n_2(n_1 + n_2 - 2)}{(n_1\bar{X} + n_2\bar{Y})(\sum_1^{n_1} X_i^{-1} + \sum_1^{n_2} Y_i^{-1}) - (n_1 - n_2)^2} \right]^{1/2},$$

$$\rho = \left[\frac{(n_1\bar{X} + n_2\bar{Y})(\sum_1^{n_1} X_i^{-1} + \sum_1^{n_2} Y_i^{-1}) - (n_1 - n_2)^2}{(n_1\bar{X} + n_2\bar{Y})(\sum_1^{n_1} X_i^{-1} + \sum_1^{n_2} Y_i^{-1}) - (n_1 + n_2)^2} \right]^{(n_1 + n_2 - 3)/2},$$

and $n = n_1 + n_2$.

In the two-sided alternative case the rejection region is given by

$$\left| \frac{[n_1 n_2(n_1 + n_2 - 2)]^{1/2}(\bar{X} - \bar{Y})}{[\bar{X}\bar{Y}(n_1\bar{X} + n_2\bar{Y})(V_1 + V_2)]} \right| > t_{1-\alpha/2}, \tag{6.31}$$

where $t_{1-\alpha/2}$ is the $100(1 - \alpha/2)$ percentage point of the Student's t distribution with $(n_1 + n_2 - 2)$ degrees of freedom.

It is interesting that when two samples are of equal size, $n_1 = n_2$, one finds from (6.27) that $C = z_{1-\alpha}$ and from (6.30),

$C = t_{1-\alpha}$. When n_1 and n_2 are moderately large, the second term in the left side of (6.27) (or 6.30) is negligible and thus the critical value $z_{1-\alpha}$ (or $t_{1-\alpha}$) provides a good approximation for C.

The UMP unbiased test in (6.31) is also the likelihood ratio test, as pointed out by Davis (1980). She also discussed the likelihood ratio test for the equality of the scale parameters λ and τ of two inverse Gaussian populations, $IG(\mu, \lambda)$ and $IG(v, \tau)$. The null test statistic is $R = (n_2 - 1)V_1/(n_1 - 1)V_2 \sim F_{n_1-1, n_2-1}$ as shown in Theorem 5.2. Moreover, it can be easily seen that the likelihood ratio test statistics for the one-sided alternative hypotheses are the same as the optimum test statistics discussed here. The null distributions of these statistics can be easily obtained from the sampling distributions discussed in Section 5.4 or the related statistical distributions given in Section 4.1. As some of these distributions would involve the nuisance parameters, an exact evaluation of a one-sided likelihood ratio test may not be possible. However, an approximation similar to that of an optimum test discussed above is applicable to the likelihood ratio test, as well.

It is desirable to investigate the power of these tests. However, the exact non-null distributions of the test statistics are difficult to evaluate. One of the reasons for such difficulty is that the additive property of the inverse Gaussian random variable is not applicable under any of the alternative hypotheses. Of course, because of the analogy that one finds between the present results and those for the normal distributions, one may be able to visualize the nature of these nonnull distributions. A general discussion on the related distributions for the inverse Gaussian given by Chhikara and Folks (1975) could lead to an understanding of these distributions.

6.3.2 Confidence Intervals

The tests in Section 6.3.1 can be extended to compare the two inverse Gaussian means in terms of their ratio. This follows because of the property that the density function $f(x; \mu, \lambda) = \mu^{-1}f(x; 1, \lambda/\mu)$ for $X \sim IG(\mu, \lambda)$. UMP unbiased tests of the hypotheses

$$H_0: \frac{\mu}{v} \leqslant \Delta \qquad \text{versus} \qquad H_A: \frac{\mu}{v} > \Delta, \qquad \Delta > 0$$

and

$$H_0 : \frac{\mu}{\nu} = \Delta \qquad \text{versus} \qquad H_A : \frac{\mu}{\nu} \neq \Delta, \qquad \Delta > 0 \qquad (6.32)$$

can be derived from those given in (6.26) and (6.28) when λ is assumed known and from those given in (6.29) and (6.31) when λ is unknown, respectively, by replacing Y_i by ΔY_i, $i = 1, 2, \ldots, n_2$. This is of course feasible when the scale parameter of the distribution function of Y_i is assumed to be $\lambda \Delta^{-1}$. Under this assumption the additive property of the inverse Gaussian holds for the combined samples $X_1, \ldots, X_{n_1}$ and $Y_1, Y_2, \ldots, Y_{n_2}$, and the null distributions of the resulting test statistics can be derived in the same manner as they were for the case of testing the equality of two means.

It is straightforward to express the UMP-unbiased test procedures in terms of Δ for hypotheses in (6.32); we omit their exact expression. Instead, we give here the $100(1 - \alpha)$ percent confidence intervals for Δ obtained by inverting the acceptance regions of these tests at level α.

When λ is *known*, the two-sided interval of Δ is

$$(a[b - cz_{1-\alpha/2}], a[b + cz_{1-\alpha/2}]), \qquad (6.33)$$

where

$$a = \frac{\bar{X}}{\bar{Y}} \left(1 - \frac{\bar{X}}{n_1 \lambda} z_{1-\alpha/2}^2 \right)^{-1}$$

$$b = 1 + \frac{\bar{X}}{2n_2 \lambda} z_{1-\alpha/2}^2$$

$$c = \left[\frac{(n_1 + n_2)\bar{X}}{n_1 n_2 \lambda} + \left(\frac{\bar{X}^2}{4n_2^2 \lambda^2} \right) z_{1-\alpha/2}^2 \right]^{1/2}.$$

The two-sided interval in (6.33) is obtained provided $(\bar{X}/(n_1\lambda))z_{1-\alpha/2}^2 < 1$. Otherwise, the interval is one-sided and is given by

$$(0, a'[-b + cz_{1-\alpha/2}]), \qquad (6.34)$$

where

$$a' = \frac{\bar{X}}{\bar{Y}}\left(-1 + \frac{\bar{X}}{n_1\lambda}z_{1-\alpha/2}^2\right)^{-1}.$$

When λ *is unknown*, the confidence interval for Δ is similar to (6.33) or (6.34) and is given by

$$(A[B - Ct_{1-\alpha/2}], A[B + Ct_{1-\alpha/2}]), \tag{6.35}$$

where

$$A = \frac{\bar{X}}{\bar{Y}}\left(1 - \frac{\bar{X}V_1}{n_1}d^2\right)^{-1},$$

$$B = 1 + \left(\frac{1}{2}\right)\left(\frac{\bar{X}V_1}{n_2} + \frac{\bar{Y}V_2}{n_1}\right)d^2,$$

$$C = d\left[\left(\frac{\bar{X}}{n_1} + \frac{1}{n_2\bar{Y}}\right)V_1 + \left(\frac{\bar{Y}}{n_2} + \frac{1}{n_1\bar{Y}}\right)V_2\right.$$
$$\left. + \left(\frac{1}{4}\right)\left(\frac{\bar{X}V_1}{n_2} - \frac{\bar{Y}V_2}{n_1}\right)^2 d^2\right]^{1/2},$$

$$d = (n_1 + n_2 - 2)^{-1/2}t_{1-\alpha/2}$$

and V_1 and V_2 are as defined in Section 6.3.1. Once again, the interval in (6.35) is obtained provided $1 - \bar{X}V_1d^2/n_1 > 0$. Otherwise, it is

$$(0, A'[-B + Ct_{1-\alpha/2}]), \tag{6.36}$$

where

$$A' = \frac{\bar{X}}{\bar{Y}}\left(-1 + \frac{\bar{X}V_1}{n_1}d_1^2\right)^{-1}.$$

Next, an exact $100(1 - \alpha)$ percent lower confidence limit of Δ cannot be given in an explicit form. However, as in the one-sample case these can be approximated using $z_{1-\alpha}$, the $100(1 - \alpha)$

percentage point of the standard normal, for λ known and using $t_{1-\alpha}$, the $100(1-\alpha)$ percentage point of the Student's t distribution with $(n_1 + n_2 - 2)$ degrees of freedom, for λ unknown.

The exact confidence limits presented above are fairly complex. We illustrate these limits numerically for the following example of failure data.

Example

Gacula and Kubala (1975) reported certain sensory failure data for two refrigerated food products, M and K as these were called, and studied their shelf life by considering the Weibull and the lognormal distributions as shelf lifetime models. Both distributions provided good fits for the failure data, which are given and summarized in our context as follows:

Product M:
Shelf life in days (x):
24, 24, 26, 26, 32, 32, 33, 33, 33, 35, 41, 42, 43, 47, 48, 48, 48, 50, 52, 54, 55, 57, 57, 57, 57, 61

$$n_1 = 26, \qquad \sum x_i = 1115, \qquad \sum x_i^{-1} = .660$$
$$\bar{x} = 42.885, \qquad v_1 = \sum x_i^{-1} - \frac{n_1}{\bar{x}} = .0537.$$

Product K:
Shelf life in days (y):
21, 23, 25, 38, 43, 43, 52, 56, 61, 63, 67, 69, 70, 75, 86, 107

$$n_2 = 17, \qquad \sum y_i = 968, \qquad \sum y_i^{-1} = .3658$$

$$\bar{y} = 56.941, \qquad v_2 = \sum y_i^{-1} - \frac{n_2}{\bar{y}} = .0672.$$

The data for each product fit the inverse Gaussian distribution well. For example, the Kolmogorov-Smirnov statistic $D_n = .1378$ when the product M data are fitted (Folks and Chhikara, 1978). Next, the assumption about the scale parameters, $\lambda_Y = \lambda_X \Delta^{-1}$, seems to hold true. It then follows from (6.35) that a 95 percent confidence interval for the ratio of their means is (.635, .905), thereby rejecting the hypothesis of equal means for the two refrigerated food products.

6.3.3   Comparison of Two Inverse Gaussian Populations with Equal Means

The two-sample theory results of Section 5.4 provide a basis for making inferences about the difference between two IG popula- tions, say $IG(\mu, \lambda_i)$, $i = 1, 2$. Clearly, one can consider the variable R or Q of Theorem 5.2 for a test statistic and obtain a significance test as well as construct a confidence interval for λ_1/λ_2. Derivations are straightforward and hence are omitted.

6.4   ANALYSIS OF RESIDUALS

6.4.1   One-way Classification

Given samples of size n_i from $IG(\mu_i, \lambda_i)$, $i = 1, 2, \ldots, k$, we consider two standard hypotheses of the k-sample problem:

1.   All have the same means, given that they all have the same λ.
2.   All populations have the same λ with different, unspecified means, μ_i.

We first address the problem for case 1. Suppose X_{i1}, $X_{i2}, \ldots, X_{in_i}$ are random observations from population i, and $n = \sum n_i$. Then the maximum likelihood estimates of parameters μ and λ, where $\mu = \mu_1 = \mu_2 = \cdots = \mu_k$ are $\hat{\mu} = \bar{X}$ and

$$\frac{1}{\hat{\lambda}} = \frac{1}{n} \sum_i \sum_j \left(\frac{1}{X_{ij}} - \frac{1}{\bar{X}} \right),$$

where $\bar{X} = \sum_i \sum_j X_{ij}/n$ and $\bar{X}_i = \sum_j X_{ij}/n_i$.

The maximum of the likelihood function for the restricted situation of equal μ_i's, denoted by L_ω, is

$$L_\omega(\hat{\mu}, \hat{\lambda}) = \left(\frac{1}{2\pi} \right)^{n/2} \left(\prod_i \prod_j X_{ij}^{-3/2} \right) \left[\frac{1}{n} \sum_i \sum_j \left(\frac{1}{X_{ij}} - \frac{1}{\bar{X}} \right) \right]^{-n/2} e^{-n/2}.$$

When the μ_i are not restricted by the hypothesis, the maximum likelihood estimates of parameters $\mu_1, \mu_2, \ldots, \mu_k$ and λ are easily obtained as

$$\hat{\mu}_i = \bar{X}_i, \qquad i = 2, \ldots, k$$

$$\frac{1}{\hat{\lambda}} = \frac{1}{n} \sum_i \sum_j \left(\frac{1}{X_{ij}} - \frac{1}{\bar{X}_i} \right).$$

The unrestricted likelihood function corresponding to $\hat{\mu}_i$ and $\hat{\lambda}$, denoted by L_Ω, is

$$L_\Omega(\hat{\mu}_i\text{'s}, \hat{\lambda}) = (2\pi e)^{-n/2} \left(\prod_i \prod_j X_{ij}^{-3/2} \right)$$

$$\times \left[\frac{1}{n} \sum_i \sum_j \left(\frac{1}{X_{ij}} - \frac{1}{\bar{X}_i} \right) \right]^{-n/2}.$$

Hence the likelihood ratio $L = L_\omega / L_\Omega$ leads to

$$L^{2/n} = \frac{\sum_i \sum_j (1/X_{ij} - 1/\bar{X}_i)}{\sum_i \sum_j (1/X_{ij} - 1/\bar{X})}. \tag{6.37}$$

The denominator can be decomposed into the sum of two independently distributed χ^2 variates, that is,

$$\sum_i \sum_j \left(\frac{1}{X_{ij}} - \frac{1}{\bar{X}} \right) = \sum_i \sum_j \left(\frac{1}{X_{ij}} - \frac{1}{\bar{X}_i} \right) + \sum_i n_i \left(\frac{1}{\bar{X}_i} - \frac{1}{\bar{X}} \right), \tag{6.38}$$

where

$$\sum_i \sum_j \left(\frac{1}{X_{ij}} - \frac{1}{\bar{X}_i} \right) \sim \left(\frac{1}{\lambda} \right) \chi^2_{n-k}$$

and

$$\sum_i n_i \left(\frac{1}{\bar{X}_i} - \frac{1}{\bar{X}} \right) \sim \left(\frac{1}{\lambda} \right) \chi^2_{k-1}.$$

One of these variates is the same as the component in the

numerator. It follows that the likelihood ratio test statistic

$$W = \frac{\sum_i n_i (1/\bar{X}_i - 1/\bar{X})/(k-1)}{\sum_i \sum_j (1/X_{ij} - 1/\bar{X}_i)/(n-k)} \sim F_{k-1,n-k}. \qquad (6.39)$$

Thus we have an F-test for testing the equality of k IG population means. The level α rejection region is given by

$$W > F_{k-1,n-k,1-\alpha}, \qquad (6.40)$$

where $F_{k-1,n-k,1-\alpha}$ is the $100(1-\alpha)$ percentage point of the F distribution with $(k-1)$ and $(n-k)\,df$.

For $k = 2$ the test in (6.40) is equivalent to the UMP unbiased test of Section 6.3.1.

The analogy between the test in (6.40) and the usual test of equality of means in a one-way classification of normal populations is striking. Since the statistic W and the decomposition in (6.38) are expressed in terms of reciprocals of the sample observations X_{ij}, and because of the analogy with analysis of variance, Tweedie (1957a) called this the *analysis of reciprocals*.

To illustrate the analysis of reciprocals and compare it with the standard analysis of variance, consider the following data given in McCool (1979).

Example 1

Table 6.1 shows the results of an experiment designed to compare the performance of high-speed turbine bearings made out of five different compounds. In the experiment 10 bearings of each type were tested and the failure times in units of millions of cycles were recorded.

The analysis of variance (ANOVA) and the analysis of reciprocals (ANORE) of the failure-time observations are given in Table 6.2. Here SR stands for the "sum of deviations of reciprocals" as discussed above and MR = SR/df.

Since $F_{4,45,.05} = 2.59$, both ANOVA and ANORE declare a significant difference between the mean failure times of five compounds. Thus the hypothesis that all compounds have the same mean failure time is rejected under both types of analysis.

Table 6.1 Failure Times of Bearing Specimens

I	II	III	IV	V
3.03	3.19	3.46	5.88	6.43
5.53	4.26	5.22	6.74	9.97
5.60	4.47	5.69	6.90	10.39
9.30	4.53	6.54	6.98	13.55
9.92	4.67	9.16	7.21	14.45
12.51	4.69	9.40	8.14	14.72
12.95	5.78	10.19	8.59	16.81
15.21	6.79	10.71	9.80	18.39
16.04	9.37	12.58	12.28	20.84
16.84	12.75	13.41	25.46	21.51

Source: McCool, 1979.

Table 6.2a Analysis of Variance (ANOVA) for the Failure Times

Source of variation	df	SS	MS	F
Between compounds	4	401.28	100.32	5.02
Within compounds	45	899.24	19.98	
Total	49	1300.51		

Table 6.2b Analysis of Reciprocals (ANORE) for the Failure Times

Source of variation	df	SR	MR	F
Between components	4	0.435	0.109	4.57
Within components	45	1.070	0.024	
Total	49	1.505		

The non-null distribution of statistic W is difficult to evaluate and so the power of the likelihood ratio test has not been investigated. Miura (1978) reports the power of this test estimated for a few cases by simulation.

Next we consider case 2, the hypothesis that all populations have the same λ with different, unspecified means μ_i and give the likelihood ratio test.

Denote the maximum likelihood subject to $\lambda_1 = \lambda_2 = \cdots = \lambda_k$ by L_ω and the maximum likelihood without restrictions by L_Ω. From the work already presented, it should be clear that the maximum likelihood estimates are as follows:

Unrestricted: $\hat{\mu}_i = \bar{X}_i$

$\hat{\lambda}_i = n_i / \sum_j (1/X_{ij} - 1/\bar{X}_i)$

Restricted: $\hat{\mu}_i = \bar{X}_i$

$\tilde{\lambda} = n / \sum_i \sum_j (1/X_{ij} - 1/\bar{X}_i)$

where $\bar{X}_i = \sum_j X_{ij}/n_i$ and $n = \sum_i n_i$.

Substituting the maximum likelihood estimates into the likelihood function, we obtain

$$L_\omega = \tilde{\lambda}^{n/2} \prod_{i=1}^{k} \prod_{j=1}^{n_i} (2\pi X_{ij}^3)^{-(1/2)} e^{-n/2} \qquad (6.42)$$

$$L_\Omega = \prod_{i=1}^{k} \hat{\lambda}_i^{n_i/2} \prod_{j=1}^{n_i} (2\pi X_{ij}^3)^{-(1/2)} e^{-n/2} \qquad (6.42)$$

So we obtain the likelihood ratio,

$$LR = \frac{\tilde{\lambda}^{n/2}}{\prod_1^k \hat{\lambda}_i^{n_i/2}} . \qquad (6.43)$$

This result is strikingly similar to the likelihood ratio for the equality of variances when sampling from the normal distribution. Let $V_i = \sum_j (1/X_{ij} - 1/\bar{X}_i)$ and $V = \sum V_i$ so that the likelihood ratio is

$$LR = \frac{\prod_{i=1}^{k} (V_i/n_i)^{n_i/2}}{(V/n)^{n/2}} . \qquad (6.44)$$

The resemblance to the normal test becomes even more striking. In general, we know that $-2 \ln (LR)$ is approximately a chi-square variable under H_0 and we can make use of that approximation here. Also under H_0, $\lambda V_i \sim \chi^2_{n_i-1}$; so Bartlett's approximation (1937) holds. This requires that all of the n_i represented in equation (6.44) be replaced by degrees of freedom $f_i = n_i - 1$ and that the resulting statistic be divided by

$$C = 1 + \frac{1}{3(k-1)}\left[\sum\frac{1}{f_i} - \frac{1}{\sum f_i}\right]. \tag{6.45}$$

The modified test statistic is M/C where

$$M = f \ln\left(\frac{V}{f}\right) - \sum f_i \ln\left(\frac{V_i}{f_i}\right) \tag{6.46}$$

with $f = \sum(n_i - 1)$ and C as given by (6.45). Under H_0, M/C is distributed approximately as chi-square with $(k - 1)$ degrees of freedom.

Example 2

Consider the failure time data given in Table 6.1. Suppose one wants to test the hypothesis that the λ_i are equal. The computed value of the test statistic M/C is easily obtained.

$$M = 45 \ln\left(\frac{1.0703}{45}\right) - 9\sum_1^5 \ln\left(\frac{V_i}{9}\right)$$

$$= 3.5354$$

$$C = 1.0444$$

and

$$\frac{M}{C} = 3.385$$

Since $\chi^2_{4,.05} = 9.49$ exceeds the observed value, 3.385, the null hypothesis of equal λ_i is not rejected at the .05 significance level.

6.4.2   Two-Way Classification

The analogy with the analysis of variance seen earlier is restricted to the nested classification. Suppose we consider a cross-classification model with the random variable X_{ij} from row-column cell (i, j) distributed as $\text{IG}(\mu + \alpha_i + \beta_j + \delta_{ij}, \lambda)$. Then the algebraic identity for the main effects and row-column interaction in the analysis of reciprocals,

$$\sum_i \sum_j \left(\frac{1}{X_{ij}} - \frac{1}{\overline{X}} \right) = \sum_i \sum_j \left(\frac{1}{\overline{X}_{i.}} - \frac{1}{\overline{X}} \right) + \sum_i \sum_j \left(\frac{1}{\overline{X}_{.j}} - \frac{1}{\overline{X}} \right)$$

$$+ \sum_i \sum_j \left(\frac{1}{X_{ij}} - \frac{1}{\overline{X}_{i.}} - \frac{1}{\overline{X}_{.j}} + \frac{1}{\overline{X}} \right) \qquad (6.47)$$

does not give independent components. The interaction term

$$\sum_i \sum_j \left(\frac{1}{X_{ij}} - \frac{1}{\overline{X}_{i.}} - \frac{1}{\overline{X}_{.j}} + \frac{1}{\overline{X}} \right)$$

can be negative and does not have a distribution of the chi-square type (Tweedie, 1957a).

Though we lack here a complete analogy with the analysis of variance, tests for main effects can be obtained by considering a nested classification for row (column) totals obtained by summing over columns (rows). These tests for the two-way classification were given by Shuster and Miura (1972) for equal subclass numbers and by Miura (1978) for the unequal case. Though a test of interaction in the case of equal subclass numbers was developed by Shuster and Miura (1972), it suffered from the requirement of many observations in each cell.

We discuss the likelihood ratio test for row effects in n_{ij} observations for cell (i, j), $i = 1, 2, \ldots, R$ and $j = 1, 2, \ldots, C$. Denote $n_{i.} = \sum_j n_{ij}$, $n_{.j} = \sum_i n_{ij}$ and $n_{..} = \sum_i \sum_j n_{ij}$. Let $X_{ijk} \sim \text{IG}(\mu + \alpha_i + \beta_j, \lambda)$ where $\mu > 0$, $\mu + \alpha_i + \beta_j > 0$, $\sum_i \alpha_i = 0$, and $\sum_j \beta_j = 0$. The following hypothesis is to be tested:

$$H_0: \alpha_1 = \alpha_2 = \cdots = \alpha_R \qquad (6.48)$$

against

H_A: Not all α_i are equal

The maximum likelihood estimates are straightforward to obtain under H_0 (i.e., the restricted case) as well as under H_A (i.e., the unrestricted case).

Evaluating the maximum of the likelihood function in each case and considering their ratio, the following statistic is obtained:

$$T = \frac{\sum_i \sum_j \sum_k (1/X_{ijk} - 1/\bar{X}_{ij.})}{\sum_i \sum_j \sum_k (1/X_{ijk} - 1/\bar{X}_{.j.})},$$

where

$$\bar{X}_{ij.} = \frac{1}{n_{ij}} \sum_k X_{ijk}$$

$$\bar{X}_{.j.} = \frac{1}{\sum_i n_{ij}} \sum_i \sum_k X_{ijk}.$$

The term in the denominator of T is the sum of the term in the numerator which is distributed as $(1/\lambda)\chi^2_{n-RC}$, and another independent term,

$$\sum_i \sum_j n_{ij} \left(\frac{1}{\bar{X}_{ij.}} - \frac{1}{\bar{X}_{.j.}} \right) \sim \left(\frac{1}{\lambda} \right) \chi^2_{C(R-1)}.$$

Thus T is distributed as a constant times a beta random variable and can be transformed to the statistic,

$$S = \frac{\sum_i \sum_j n_{ij}(1/\bar{X}_{ij.} - 1/\bar{X}_{.j.})/C(R-1)}{\sum_i \sum_j \sum_k (1/X_{ijk} - 1/\bar{X}_{ij.})/\sum_i \sum_j (n_{ij} - 1)} \tag{6.49}$$

which has the F distribution with $C(R-1)$ and $\sum_i \sum_j (n_{ij} - 1)$ degrees of freedom. Hence the F-test is obtained with the test statistic S to test the hypothesis in (6.48).

The test for the hypothesis H_0: all β_j are equal against H_A: not all β_j are equal is analogous to the above test and will not be discussed here.

Based on a limited simulation study Miura (1978) compared the power of the above test with that of the usual analysis of variance F-test. She also considered another F-test (named VF) obtained after applying the variance-stabilizing transformation $t(x) = x^{-1/2}$ to the inverse Gaussian observations. She concluded the present test was preferred over the F and VF tests since its power values were superior overall.

Fries and Bhattacharyya (1983) proposed to model the reciprocal of the cell mean in two-factor experiments involving failure times distributed as inverse Gaussian. If μ_{ij} denotes the mean for cell (i, j), a no-interaction model was assumed for μ_{ij}^{-1} as follows:

$$\mu_{ij}^{-1} = \theta + \gamma_i + \delta_j, \qquad \sum \gamma_i = \sum \delta_j = 0 \tag{6.50}$$

where θ, γ_i's and δ_j's represent the overall mean, the row effects, and the column effects, respectively. Based on the reciprocals of individual observations and also of their cell, row, and column means, an analysis of reciprocals (ANORE) was developed. Although the authors showed ANORE having some similarity with the usual ANOVA for the two-way classification, in fact, there is no parallel when the analysis is interpreted in terms of the assumed model. This is because the ANORE includes a two-factor "interaction" term whereas the model in (6.50) does not. We wonder how one can make such analysis more meaningful.

7

Bayesian Inference

7.1 ON THE CHOICE OF A PRIOR DISTRIBUTION

In Bayesian inference for the inverse Gaussian the parameterization affects the choice of prior distributions. For some parameterizations a natural conjugate prior distribution exists; for others it does not. Reparameterization merely to achieve a natural conjugate prior may seem artificial; on the other hand, when one parameterization is as reasonable as another, it offers no philosophical difficulties. In the following sections we shall consider for certain parameterizations the natural conjugate prior and some diffuse priors.

7.2 THE PARAMETERIZATION (μ, λ)

7.2.1 The Natural Conjugate Prior

We have seen earlier that the (μ, λ) parameterization lends itself to a useful statistical methodology of parametric estimation and hypothesis testing, and thus it is desirable, if possible, to stay with this parameterization. Palmer (1973) noted that the natural conjugate prior does not exist for this parameterization except when μ is known and λ is unknown.

The density function can be written in the form:

$$f(x; \mu, \lambda) = (2\pi x^3)^{-1/2} \left[\lambda^{1/2} \exp\left(-\frac{\lambda x}{2\mu^2} + \frac{\lambda}{\mu} - \frac{\lambda}{2x} \right) \right]. \quad (7.1)$$

The kernel of $f(x; \mu, \lambda)$ is the term within brackets. If a natural conjugate prior existed, it would have the same form as the kernel so that the form of the prior density function would be

$$p(\mu, \lambda) \propto \lambda^d \exp[-\lambda(a\mu^{-2} - b\mu^{-1} + c)],$$

$$0 < \mu < \infty, 0 < \lambda < \infty. \quad (7.2)$$

Since $(x - \mu)^2/2\mu^2 x > 0$ it is necessary that $a\mu^{-2} - b\mu^{-1} + c > 0$; in other words, that $a > 0$ and $b^2 - 4ac < 0$. In addition, the nature of the gamma function requires that $d > -1$ in order that $p(\mu, \lambda)$ be λ-integrable. The corresponding posterior distribution would then have the same form as Equation (7.2) with

$$a' = a + \frac{n\bar{x}}{2}, \qquad\qquad b' = b + n$$

$$c' = c + \frac{1}{2}\sum x_i^{-1}, \qquad d' = d + \frac{n}{2}. \quad (7.3)$$

To see that the natural conjugate prior does not always exist, consider λ fixed and let $\mu \to \infty$. From equation (7.2) we can see that $\lim_{\mu \to \infty} p(\mu, \lambda) = $ constant. Therefore $p(\mu, \lambda)$ is not μ-integrable for any λ and the natural conjugate prior does not exist.

With λ known the functional form of the conjugate prior, if it exists, is

$$p(\mu) \propto \exp[-(a\mu^{-2} - b\mu^{-1})], \qquad \mu > 0, a > 0, b > 0. \quad (7.4)$$

However, as in the case with both parameters unknown, this is not a proper density because it is not integrable.

With μ known the natural conjugate for λ is of the form

$$p(\lambda) \propto \lambda^a \exp(-b\lambda), \qquad \lambda > 0, b > 0. \quad (7.5)$$

Since this is the form of a gamma, take

$$p(\lambda) = \frac{b^a \lambda^{a-1} e^{-b\lambda}}{\Gamma(a)}, \qquad a > 0, b > 0, \lambda > 0. \tag{7.6}$$

The posterior distribution of λ is then a gamma distribution with parameters

$$a' = a + \frac{n}{2} \qquad \text{and} \qquad b' = b + \sum \frac{(x_i - \mu)^2}{2\mu^2 x_i}.$$

7.2.2 Jeffrey's Prior Distribution

First we need to obtain the Fisher information matrix. The derivatives of $f(x; \mu, \lambda)$ with respect to μ and λ are

$$\frac{\partial \log f}{\partial \lambda} = \frac{1}{2\lambda} - \frac{1}{2x} \frac{(x - \mu)^2}{\mu^2}, \qquad \frac{\partial^2 \log f}{\partial \lambda^2} = -\frac{1}{2\lambda^2},$$

$$\frac{\partial \log f}{\partial \mu} = \frac{\lambda(x - \mu)}{\mu^3}, \qquad \frac{\partial^2 \log f}{\partial \mu^2} = -\frac{\lambda(3x - 2\mu)}{\mu^4},$$

$$\frac{\partial^2 \log f}{\partial \mu \partial \lambda} = \frac{x - \mu}{\mu^3}.$$

Now the Fisher information matrix is easily seen to be

$$I(\mu, \lambda) = \begin{bmatrix} (2\lambda^2)^{-1} & 0 \\ 0 & \lambda \mu^{-3} \end{bmatrix}. \tag{7.8}$$

Since Jeffrey's prior is proportional to $\{\det I(\mu, \lambda)\}^{1/2}$, the density function is

$$p_J(\mu, \lambda) \propto \lambda^{-1/2} \mu^{-3/2}, \qquad \mu > 0, \lambda > 0. \tag{7.9}$$

This distribution is obviously an improper distribution.

To obtain the posterior distribution, we multiply the prior

distribution by the likelihood to obtain

$$p(\mu, \lambda \mid \mathbf{x}) \propto \lambda^{(n-1)/2} \mu^{-3/2}$$

$$\times \exp\left[-\lambda\left(\frac{\sum x_i}{2}\mu^{-2} - n\mu^{-1} + \frac{\sum x_i^{-1}}{2}\right)\right], \qquad \mu > 0, \lambda > 0.$$

$$(7.10)$$

Palmer (1973) shows that expression (7.10) can be normed and therefore that the posterior density is a proper density.

With μ known, Jeffrey's prior for λ is proportional to $\lambda^{-(1/2)}$ and therefore the posterior distribution of λ is a gamma distribution.

When λ known, Jeffrey's prior for μ is proportional to $\mu^{-(3/2)}$ and the posterior has the form

$$p(\mu) \propto \mu^{-3/2} \exp(-a\mu^{-2} + b\mu^{-1}),$$

$$\mu > 0, a > 0, b > 0. \qquad (7.11)$$

This has the same form as (7.10) and is therefore a proper density.

We can use the posterior distributions obtained from Jeffrey's prior to obtain various estimates of μ and λ. The marginal posterior distribution of μ is obtained from equation (7.10) to be

$$\int_0^\infty p(\mu, \lambda \mid \mathbf{x})\, d\lambda \propto \frac{\Gamma[(n+1)/2]\mu^{n-1/2}}{(a - n\mu + b\mu^2)^{(n+1)/2}}, \qquad \mu > 0, \qquad (7.12)$$

where $a = \sum x_i/2$ and $b = \sum x_i^{-1}/2$.

Although the marginal posterior distribution of μ is a proper distribution, the first moment (and consequently, higher moments) does not exist. If the estimator of μ is to be based on the posterior distribution, we will have to use either the median or the mode of the posterior distribution. Although no explicit algebraic expression for the median has been found, Palmer does give the modal value as

$$\mu_{\text{mode}} = \frac{n-2}{3}\left\{-\bar{x}_h + \left[\bar{x}_h^2 + \frac{3(2n-1)\bar{x}\bar{x}_h}{(n-2)^2}\right]^{1/2}\right\}, \qquad (7.13)$$

where $\bar{x}$ and $\bar{x}_h$ are the arithmetic and harmonic sample means, respectively.

When λ is known, the posterior distribution of μ is proper but again the first moment does not exist. However, the modal value has the simple form:

$$\mu_{\text{mode}} = \frac{2\bar{x}}{1 + [1 + 6\bar{x}/n]^{1/2}}. \tag{7.14}$$

The marginal posterior distribution of λ offers no problems conceptually but so far has not proven tractable. On the other hand, the posterior distribution of λ for μ known is a gamma distribution such that

$$K\lambda \sim \chi_n^2, \tag{7.15}$$

where

$$K = \left(\frac{\sum x_i}{2\mu^2} - \frac{n}{\mu} + \frac{\sum x_i^{-1}}{2}\right).$$

Making use of the chi-square distribution, we can estimate the posterior mean, median, mode, and so on.

Instead of using the marginal posterior distributions of μ and λ we might choose to use some other parametric function of the joint posterior distribution. Actually, the modal value of $p(\mu, \lambda)$ is tractable and fairly easily obtained from equation (7.10).

7.3 THE PARAMETERIZATION $(1/\mu, \lambda)$

7.3.1 Introduction

In quite a number of cases, as Tweedie noticed originally, results come easier if we work with the reciprocal of the mean. Let $\theta = 1/\mu$ and express the density in terms of θ and λ. Making use of

$$\frac{\lambda}{2\mu^2} \sum \frac{(x_i - \mu)^2}{x_i} = \frac{\lambda}{2} \sum \left(\frac{1}{x_i} - \frac{1}{\bar{x}}\right) + \frac{n\lambda(\bar{x} - \mu)^2}{2\mu^2 \bar{x}}, \tag{7.16}$$

we can write the joint density function of X_i's as

$$f(x; \theta, \lambda) = \left(\frac{\lambda}{2\pi}\right)^{n/2} \prod_1^n x_i^{-3/2}$$

$$\times \exp\left[-\frac{\lambda v}{2} - \frac{n\lambda \bar{x}}{2}\left(\theta - \frac{1}{\bar{x}}\right)^2\right\}, \qquad (7.17)$$

where $v = \sum (1/x_i - 1/\bar{x})$.

7.3.2 The Natural Conjugate Prior

From (7.17) we can see that the natural conjugate prior must have the form

$$p(\theta, \lambda) \propto \lambda^a \exp\left[-\frac{a\lambda}{b}(\theta - b)^2 - c\lambda\right], \qquad (7.18)$$

where $a > 0$, $b > 0$, and $c > 0$. This prior distribution is recognized as the product of a truncated normal and a gamma-type distribution. The distribution of θ with λ specified is a normal distribution truncated at zero and the marginal of λ is a gamma distribution. Combining equations (7.17) and (7.18) we can see that the posterior distribution of (θ, λ) has the same form as the prior in (7.18) with parameters modified by $\bar{x}$ and v. Its derivation is straightforward and is given by Banerjee and Bhattacharyya (1979).

Much simpler results are obtained when we take either θ or λ to be known. With λ known, we can see from equation (7.18) that the natural conjugate prior should have the form

$$p(\theta) \propto \exp\left[-\frac{(\theta - c)^2}{2d}\right], \qquad \theta > 0. \qquad (7.19)$$

This is obtained by truncating a normal distribution with mean c and variance d. Of course, the posterior distribution has the same form and admits the usual estimates based on a normal posterior distribution.

With θ known, the natural conjugate prior distribution of λ is seen to be a gamma distribution and the posterior has the same form.

It is interesting that the natural conjugate prior exists (is a proper distribution) for the parameterization $(1/\mu, \lambda)$ but not for the parameterization (μ, λ). This points up once again the heavy dependence of the Bayesian approach upon the choice of parameters.

7.3.3 Jeffrey's Prior Distribution

With the parameterization (θ, λ) we can obtain Jeffrey's prior distribution. The Fisher information matrix for a sample of size n is

$$I(\theta, \lambda) = \begin{bmatrix} n\lambda/\theta & 0 \\ 0 & n/(2\lambda^2) \end{bmatrix}, \tag{7.20}$$

so Jeffrey's prior distribution has the form

$$p(\theta, \lambda) \propto [\det I(\theta, \lambda)]^{1/2}$$

$$\propto \frac{n}{\sqrt{\theta\lambda}}. \tag{7.21}$$

In this case the use of Jeffrey's prior distribution produces a posterior distribution which is not a proper distribution. Because of this, other diffuse priors have been studied.

7.3.4 Vague Prior Distribution

Banerjee and Bhattacharyya (1979) considered the locally uniform reference prior

$$p(\theta \mid \lambda) \propto \text{constant}, \qquad p(\lambda) \propto \lambda^{-1}. \tag{7.22}$$

Then the joint posterior distribution has the density function

$$p(\theta, \lambda \mid \mathbf{x}) = K\lambda^{n/2 - 1} \exp\left\{ -\frac{n\lambda}{2\hat{\lambda}} \left[1 + \frac{\hat{\lambda}}{\bar{x}} (\theta\bar{x} - 1)^2 \right] \right\}, \tag{7.23}$$

where

$$\frac{1}{\hat{\lambda}} = \frac{1}{n}\left[\sum \frac{1}{x_i} - \frac{n}{\bar{x}}\right], \qquad \bar{x} = \frac{1}{n}\sum x_i.$$

and K is the normalizing constant. A contour of constant value for $p(\theta, \lambda \mid \mathbf{x})$ is given by

$$\left(\frac{n}{2} - 1\right)\log \lambda - \frac{n\lambda}{2\hat{\lambda}}\left[1 + \frac{\hat{\lambda}}{\bar{x}}(\theta\bar{x} - 1)^2\right] = c. \tag{7.24}$$

The mode of $p(\theta, \lambda \mid \mathbf{x})$ is $(1/\bar{x}, (n-2)\hat{\lambda}/n)$ where the left-hand side of equation (7.24) reduces to, c_0 say, so that

$$(n - 2)(\log \hat{\lambda} - 1) = 2c_0.$$

When n is large, the distribution of

$$-2[\log p(\theta, \lambda \mid \mathbf{x}) - \log p(\hat{\theta}, \hat{\lambda} \mid \mathbf{x})],$$

where $\hat{\theta} = 1/\bar{x}$, is approximately the chi-square distribution with two degrees of freedom. Based on this approximation, the contour containing approximately $(1 - \alpha)$ posterior probability is the set of points given by

$$(n - 2)\log \lambda - \frac{n\lambda}{\hat{\lambda}}\left[1 + \frac{\hat{\lambda}}{\bar{x}}(\theta\bar{x} - 1)^2\right] = 2c_0 - \chi^2_{2,\alpha}, \tag{7.25}$$

where $\chi^2_{2,\alpha}$ is the upper 100α percentage point of the chi-square distribution with two degrees of freedom.

Bannerjee and Bhattacharayya further obtained the marginal posterior distributions of the individual parameters θ and λ. It easily follows by integrating (7.23) with respect to λ that the marginal posterior of θ is given by

$$p(\theta \mid \mathbf{x}) = K\Gamma\left(\frac{n}{2}\right)\left(\frac{n}{2\hat{\lambda}}\right)^{-n/2}\left[1 + \bar{x}\hat{\lambda}\left(\theta - \frac{1}{\bar{x}}\right)^2\right]^{-n/2},$$

$$0 < \theta < \infty, \tag{7.26}$$

where

$$K = \frac{(\bar{x}\hat{\lambda})^{1/2}(n/2\hat{\lambda})^{n/2}}{S_{n-1}(\xi)B[(n-1)/2, \frac{1}{2}]\Gamma(n/2)}$$

with $S_{n-1}(\xi)$ as the cdf of Student's t distribution with $(n-1)$ degrees of freedom and $\xi = [(n-1)\hat{\lambda}/\bar{x}]^{1/2}$. Here $B(\cdot,\cdot)$ and $\Gamma(\cdot)$ are the beta and gamma functions, respectively. One may recognize that the density function in (7.26) is that of a left-truncated t distribution with $(n-1)$ degrees of freedom, location parameter $1/\bar{x}$, scale parameter $[v\bar{x}\hat{\lambda}]^{1/2}$, and the point of truncation being at zero, where $v = (n-1)$.

Next, by integrating (7.23) with respect to θ, the marginal posterior density function of λ is

$$p(\lambda|\mathbf{x}) = \frac{\Phi([n\lambda/\bar{x}]^{1/2})}{S_{n-1}(\xi)} \frac{(n\lambda/2\hat{\lambda})^{(n-1)/2}}{\lambda\Gamma((n-1)/2}$$

$$\times \exp\left(-\frac{n\lambda}{2\hat{\lambda}}\right), \qquad \lambda > 0, \tag{7.27}$$

where Φ is the standard normal cdf. It may be noted that the expression in (7.27) is a gamma pdf multiplied by a factor involving the normal and Student's t cdf's. Hence, one may call it a weighted gamma distribution. The weighting factor is

$$\frac{\Phi((n\lambda/\bar{x})^{1/2})}{S_{n-1}([(n-1)\hat{\lambda}/\bar{x}]^{1/2})},$$

which approaches 1 as $n \to \infty$. Thus, in large samples the marginal posterior distribution of λ is up to a constant approximately chi-square with $(n-1)$ degrees of freedom.

7.4 HIGHEST PROBABILITY DENSITY (HPD) REGIONS

One can obtain an interval estimate of a parameter from its posterior distribution similar to a confidence interval obtained from the sampling distribution. Often an HPD interval is desired because it is the shortest and every point inside it has probability density at least that of any point outside it. Of course, in the case of more than one parameter it will be a HPD region for the set of parameters. For definitions and further details on the Bayesian approach, one may refer to Box and Tiao (1973).

We consider the parameterization of (θ, λ) since it has proper posterior distributions that lead to explicit HPD regions for the parameters. In the case of the uniform prior we find that the marginal posterior density function given in (7.26) is symmetric over the interval $(0, 2/\bar{x})$ and has mode at $1/\bar{x}$. Accordingly, if $P[\theta \in (1/\bar{x} - a, \ 1/\bar{x} + a)|\mathbf{x}] = 1 - \alpha$, we have $a = (\bar{x}\lambda v)^{1/2} S_{n-1}^{-1}([1 + (1 - \alpha)S_{n-1}(\xi)]/2)$. Hence, the $(1 - \alpha)$HPD interval for θ is given by

$$
\frac{1}{\bar{x}} \pm (\bar{x}\hat{\lambda}v)^{1/2} S_{n-1}^{-1} \left(\frac{1 + (1 - \alpha)S_{n-1}(\xi)}{2} \right)
\tag{7.28}
$$

provided $1/\bar{x} \geqslant a$. Otherwise, the $(1 - \alpha)$HPD can be obtained by considering $P[\theta \in (0, 1/\bar{x} + b)|\mathbf{x}] = 1 - \alpha$. Again it follows from (7.26) that $b = (\bar{x}\hat{\lambda}v)^{1/2} S_{n-1}^{-1}(1 - \alpha S_{n-1}(\xi))$. Thus the $(1 - \alpha)$HPD interval of θ is

$$
\left(0, \frac{1}{\bar{x}} + (\bar{x}\hat{\lambda}v)^{1/2} S_{n-1}^{-1}(1 - \alpha S_{n-1}(\xi)) \right),
\tag{7.29}
$$

where $\xi = [(n - 1)\hat{\lambda}/\bar{x}]^{1/2}$ and $v = (n - 1)$.

The intervals in (7.28) and (7.29) are similar in form to the uniformly most accurate unbiased confidence intervals given in (6.21) for μ. This might be expected because of the vague prior used to obtain the posterior distribution in (7.23) and hence the marginal posterior in (7.26).

The marginal posterior density of λ given in (7.27) is not integrable in an exact form and hence no explicit expression for an HPD interval of λ is possible. However, a numerical solution can be obtained for the $(1 - \alpha)$HPD interval, (λ_L, λ_U) (say), where λ_L and λ_U are determined so that

$$P[\lambda \in (\lambda_L, \lambda_U) | \mathbf{x}] = 1 - \alpha \tag{7.30}$$

and

$$p(\lambda_L | \mathbf{x}) = p(\lambda_U | \mathbf{x}).$$

Since $p(\lambda | \mathbf{x})$ is unimodal, the solution of equation (7.30) is unique (Banerjee and Bhattacharyya, 1979). Of course, in large samples, the distribution of λ is approximately $(\hat{\lambda}/n)\chi^2_{n-1}$ and so the $(1 - \alpha)$HPD interval of λ is given by

$$\left[\left(\frac{\hat{\lambda}}{n} \right) \chi^2_{n-1,\alpha/2}, \left(\frac{\hat{\lambda}}{n} \right) \chi^2_{n-1,(1-\alpha)/2} \right], \tag{7.31}$$

where $\chi^2_{n-1,\alpha/2}$ and $\chi^2_{n-1,(1-\alpha/2)}$ are the $100\alpha/2$ and $100(1 - \alpha/2)$ percentage points of the chi-square distribution with $(n - 1)$ degrees of freedom. The approximate $(1 - \alpha)$HPD interval in (7.31) is the same as the uniformly most accurate unbiased $(1 - \alpha)$ confidence interval of λ discussed in Chapter 6. Thus, as in general (Box, 1980), the use of a locally uniform reference prior distribution adds little information to that contained in the sample observations when the sample size is large.

The $(1 - \alpha)$HPD region can easily be obtained for (θ, λ), as discussed previously in Section 7.3. See the results given in (7.25).

Similarly one can obtain HPD intervals based on the posterior distributions corresponding to the conjugate prior distributions of parameters θ and λ. As is seen in Section 7.3.1, there are many similarities between the natural conjugate prior and posterior distributions of the inverse Gaussian parameters and those of the normal distribution parameters. Interested readers may see the papers by Banerjee and Bhattacharyaa (1976, 1979) for further details.

The Bayesian analysis such as the determination of HPD regions is not tractable for the other parameterization of $IG(\mu, \lambda)$. This limits the scope of Bayesian inference for the inverse Gaussian, which is usually defined in terms of these parameters.

7.5   PREDICTIVE INFERENCE

Let $\mathbf{x} = (x_1, x_2, \ldots, x_n)$ be n independent observations from $IG(\mu, \lambda)$ and Y be an additional observation to be taken from it independently of $\mathbf{x}$. If $p(\theta, \lambda \mid \mathbf{x})$ is the joint posterior density function of (θ, λ), then the predictive distribution of Y, given $\mathbf{x}$, is given by the marginal density function

$$h(y \mid \mathbf{x}) = \int f(y \mid \theta, \lambda) p(\theta, \lambda \mid \mathbf{x}) \, d\theta \, d\lambda. \tag{7.32}$$

The integrand in (7.32) is the joint density function of Y and (θ, λ), conditional on the data $\mathbf{x}$. The density function $h(y \mid \mathbf{x})$ as in (7.32) can be considered as an estimate of $f(y; \theta, \lambda)$ and is used for making predictive inferences for the inverse Gaussian distribution.

Considering the diffuse prior for the parameters, and hence the resulting posterior density function given in (7.23), it follows from (7.32) that

$$h(y \mid \mathbf{x}) = c \int_0^\infty \int_0^\infty \lambda^{(n-1)/2}$$

$$\times \exp\left[-\frac{\lambda}{2}\left(\frac{n}{\bar{\lambda}} + \frac{n(\bar{x}\theta - 1)^2}{\bar{x}} + \frac{(y\theta - 1)^2}{y} \right) \right] d\theta \, d\lambda,$$

$$\tag{7.33}$$

where

$$c = \frac{[(\bar{x}\hat{\lambda}/2)y^{-3}(n/2\hat{\lambda})^n]^{1/2}}{\pi \Gamma((n - 1/2) S_{n-1}([(n-1)\hat{\lambda}/\bar{x}]^{1/2}}.$$

The exponent term, except for the factor of $(-\lambda/2)$, of the integrand can be rewritten as

$$z + (y + n\bar{x})\left[\theta - \frac{(n+1)}{(n\bar{x} + y)} \right]^2, \tag{7.34}$$

where

$$z + (y + n\bar{x})\left[\theta - \frac{(n+1)}{(n\bar{x} + y)} \right]^2,$$

Integrating (7.33) with respect to λ yields

$$h(y \mid \mathbf{x}) = c\Gamma\left(\frac{n+1}{2}\right) 2^{(n+1)/2} I, \qquad (7.35)$$

where

$$I = \int_0^\infty \left\{ z + (n\bar{x} + y)\left[\theta - \frac{n+1}{n\bar{x} + y}\right]^2 \right\}^{-(n+1)/2} d\theta.$$

The integral I can be simplified by making the transformation

$$[n(n\bar{x} + y)z^{-1}]^{1/2}\left[\theta - \frac{n+1}{n\bar{x} + y}\right] = t.$$

As derived in Chhikara and Guttman (1982), it follows from (7.35) that

$$h(y \mid \mathbf{x}) = c_0 \left[\frac{\bar{x}\hat{\lambda}}{(n\bar{x} + y)y^3}\right]^{1/2}$$

$$\times \left[1 + \frac{(\bar{x} - y)^2\hat{\lambda}}{\bar{x}y(n\bar{x} + y)}\right]^{-n/2}, \qquad y > 0 \qquad (7.36)$$

where

$$c_0 = \frac{S_n((n+1)[n/z(n\bar{x} + y)]^{1/2})}{S_{n-1}([n-1]\hat{\lambda}/\bar{x}]^{1/2})},$$

with z as in (7.34) and $S_v(\cdot)$ denoting the cdf of a Student's t distribution with v degrees of freedom.

When one parameter is known and the other one unknown, the predictive density functions are easily obtained as given in Chhikara and Guttman (1982).

Suppose we want to construct a confidence interval for a new observation, given a set of n observations sampled from an inverse Gaussian population with pdf $f(x; \theta, \lambda)$. For the Bayesian prediction limits, this can be achieved easily by using the predictive density function $h(y|\mathbf{x})$ obtained above in (7.36). Suppose we want to construct central $100(1 - \alpha)$ percent limits, say $(l(\mathbf{x}), u(\mathbf{x}))$, so that $P[y < l(\mathbf{x})|\mathbf{x}] = P[y > u(\mathbf{x})|\mathbf{x}] = \alpha/2$. Then these two-sided prediction limits $l(\mathbf{x})$ and $u(\mathbf{x})$ are given by

$$\int_0^{l(\mathbf{x})} h(y|\mathbf{x})\,dy = \frac{\alpha}{2} \quad \text{and} \quad \int_{u(\mathbf{x})}^{\infty} h(y|\mathbf{x})\,dy = \frac{\alpha}{2}, \qquad (7.37)$$

where $h(y|\mathbf{x})$ is as in (7.36). Given any α, where $0 < \alpha < 1$, a determination of $l(\mathbf{x})$ and $u(\mathbf{x})$ will require numerical integration since (7.37) cannot be solved directly. One-sided prediction limits can similarly be obtained based on the predictive density function $h(y|\mathbf{x})$ in (7.36).

In Section 9.4 we consider prediction for the inverse Gaussian variable in the classical sense and give exact as well as certain approximate prediction limits. These limits and those obtained from (7.37) are computed using a set of observed repair times. It is seen from the results given in Table 9.2 that the Bayesian limits have the shortest prediction interval. A comparison of the four types of limits is made based on their average widths and coverage probabilities; the limits based on (7.37) tend to behave overall better than others.

7.6   ADDITIONAL REMARKS

First, when both parameters are unknown, the Bayesian estimation of the reliability function $R(x)$ given in (9.1) is analytically intractable. Padgett (1981) considers the problem assuming the mean life μ is known and provides a Bayesian point estimator of $R(x)$ under noninformative prior on λ. This estimator as well as another estimator using conjugate prior on λ are discussed by Martz and Waller (1982) in Section 9.4 of their book.

Second, Athreya (1986) observed that the family of generalized inverse Gaussian densities as marginals for the variances gives rise to a new conjugate family for the normal distribution. The popular normal-gamma family is a member of this family.

8

Regression Analysis

8.1 INTRODUCTION

In this chapter we study certain regression models for the inverse Gaussian distribution and explore the possibility of skewed data analysis in a manner similar to that permissible under the usual theory of linear models. Tweedie (1957) developed what he called *analysis of reciprocals* and thereby made a beginning in this direction. Davis (1977) studied the linear regression model problem in detail. She was able to develop the analogy to the linear model theory to an advanced state for the zero-intercept model, and showed that the analogy does not extend to the most commonly used linear model with intercept. Some results were given by her for the latter case as well as for the general linear model.

Whitmore (1986) has studied the zero-intercept class of models, which includes the models considered by Davis. Bhattacharyya and Fries (1982) developed a nonlinear model which is nonlinear for the inverse Gaussian variable but linear for its reciprocal. This model can be viewed as a natural extension to the analysis of the reciprocal model discussed earlier in Section 6.4.

8.2   SIMPLE LINEAR REGRESSION MODEL: ZERO INTERCEPT

8.2.1   Estimation

Consider problems of estimation and testing for the regression model:

$$Y_i \sim \text{IG}(\beta X_i, \lambda), \qquad i = 1, 2, \ldots, n \tag{8.1}$$

with

$$\beta > 0, \, X_i > 0, \, \lambda > 0.$$

The maximum likelihood estimates of β and λ are given by

$$\hat{\beta} = \sum \frac{Y_i}{X_i^2} \Big/ \sum \frac{1}{X_i}, \qquad \hat{\lambda} = n \Big/ \sum \left(\frac{1}{Y_i} - \frac{1}{\hat{\beta} X_i} \right). \tag{8.2}$$

Because $\hat{\beta}$ is a linear combination of inverse Gaussian variables it is natural to hope that $\hat{\beta}$ is also inverse Gaussian. Recall that if independent variables $Y_i \sim \text{IG}(\mu_i, \lambda_i)$, then $\sum c_i Y_i$ is IG distributed if and only if $\lambda_i/(c_i \mu_i^2)$ is a positive constant, $i = 1, 2, \ldots, n$. Thus

$$\hat{\beta} \sim \text{IG}\left(\beta, \lambda \sum \frac{1}{X_i} \right). \tag{8.3}$$

To find the distribution of $\hat{\lambda}$, it is convenient to use the following factorization of the term in the exponent of the joint distribution:

$$\frac{\lambda}{2\beta^2} \sum \frac{(Y_i - \beta X_i)^2}{X_i^2 Y_i} = \frac{\lambda \sum (1/X_i)(\hat{\beta} - \beta)^2}{2\beta^2 \hat{\beta}} + \frac{n\lambda}{2\hat{\lambda}}. \tag{8.4}$$

This is analogous, of course, to the factorization of the exponent of the normal into terms involving $\bar{X}$ and S^2. The term on the left of equation (8.4) is a chi-square variable with n degrees of freedom. The first term on the right is a chi-square variable with one degree of freedom. We would suspect that the second term on the right is a chi-square variable with $n - 1$ degrees of freedom. To show that this is true, in a manner similar to that in Theorem 4.7, we can consider the conditional moment generating function of $n\lambda/2\hat{\lambda}$ given $\hat{\beta}$. We quickly see that it is the same for all $\hat{\beta}$ so that $n\lambda/2\hat{\lambda}$ is independent of $\hat{\beta}$. Therefore, it also follows that $n\lambda/2\hat{\lambda}$ is chi-square with $n - 1$ degrees of freedom.

Because of the factorization exhibited by equation (8.4) $(\hat{\beta}, \hat{\lambda})$ is a complete sufficient statistic for (β, λ) and therefore tests and estimates for β and λ should be based upon $\hat{\beta}$ and $\hat{\lambda}$. For example, the minimum variance unbiased estimate (MVUE) of β is $\hat{\beta}$ and because $\hat{\beta}$ is inverse Gaussian as given in equation (8.3),

$$\text{Var}(\hat{\beta}) = \beta^3/\lambda \sum \frac{1}{X_i}. \tag{8.5}$$

An unbiased estimator of $\text{Var}(\hat{\beta})$ can be found, using the method of moments, to be

$$\hat{\text{Var}}(\hat{\beta}) = \frac{[\sum(1/X_i)]^2 \sum Y^2 - [\sum(Y_i^2/X_i^2)] \sum X_i^2}{\sum X_i^3 [\sum(1/X_i)]^3 - \sum X_i^2 [\sum(1/X_i)]^2}. \tag{8.6}$$

The UMVUE of $\hat{\lambda}$ is also easily obtained. Since $n\lambda/\hat{\lambda}$ is a chi-square, $\hat{\lambda} \sim n\lambda/\chi_{n-1}^2$. Therefore, because $E(1/\chi_v^2) = 1/(v - 2)$,

$$E(\hat{\lambda}) = \frac{n\lambda}{n - 3}. \tag{8.7}$$

It follows that the UMVUE of λ is given by $\hat{\lambda}(n - 3)/n = \tilde{\lambda}$. That is,

$$\tilde{\lambda} = (n - 3)/\sum \left(\frac{1}{Y_i} - \frac{1}{\hat{\beta}X_i} \right). \tag{8.8}$$

8.2.2 Tests of Hypotheses and Confidence Intervals

Because the two terms on the right in equation (8.4) are independent chi-square variables, the idea of an F test for hypotheses concerning β is immediate. Under the null hypothesis $H_0: \beta = \beta_0$, one has

$$\frac{\lambda \sum(1/X_i)(\hat{\beta} - \beta_0)^2}{\hat{\beta}\beta_0^2} \sim \chi_1^2 \tag{8.9}$$

and

$$\frac{n\lambda}{\hat{\lambda}} \sim \chi^2_{n-1}.$$ (8.10)

Therefore

$$\frac{(n-1)\hat{\lambda}\sum(1/X_i)(\hat{\beta}-\beta_0)^2}{\beta\beta_0^2 n} \sim F_{1,n-1}.$$ (8.11)

This F test for the null hypothesis $\beta = \beta_0$ seems quite intuitive. In fact, it is also the likelihood ratio test statistic. Although the distribution under the null hypothesis is known, the distribution under the alternative hypothesis has not been determined.

The statistic given by (8.11) also provides a means of finding a confidence region for β. A $100(1-\alpha)$ percent confidence region consists of the β's satisfying

$$\frac{(n-1)\hat{\lambda}\sum(1/X_i)(\hat{\beta}-\beta)^2}{\beta\beta^2 n} \leqslant F_{1,n-1,1-\alpha}.$$ (8.12)

This set is equivalent to the set of β's satisfying

$$\beta^2\left[(n-1)\sum\frac{1}{X_i} - F_{1,n-1,1-\alpha}\left(\hat{\beta}\sum\frac{1}{Y_i} - \sum\frac{1}{X_i}\right)\right]$$

$$+ \beta\left[-2(n-1)\hat{\beta}\sum\frac{1}{X_i}\right] + (n-1)\hat{\beta}^2\sum\frac{1}{X_i} \leqslant 0.$$ (8.13)

To find the β's satisfying (8.13) first assume that the coefficient of β^2 is positive. If the quadratic in β has real roots, then we can find an interval of β's satisfying (8.13). Under this condition, the $100(1-\alpha)$ percent confidence interval is (L, U) with

$$L = \frac{\hat{\beta}}{1 + \left[\dfrac{F_{1,n-1,1-\alpha}[\hat{\beta}\sum(1/Y_i) - \sum(1/X_i)]}{(n-1)\sum(1/X_i)}\right]^{1/2}}.$$ (8.14)

and

$$U = \frac{\hat{\beta}}{1 - \left[\dfrac{F_{1,n-1,1-\alpha}[\hat{\beta}\sum(1/Y_i) - \sum(1/X_i)]}{(n-1)\sum(1/X_i)}\right]^{1/2}}. \tag{8.15}$$

If the coefficient of β^2 in (8.13) is negative, the β's satisfying (8.13) and the constraint that $\beta > 0$ are just the β's in the interval (L, ∞).

8.2.3   Other Zero-Intercept Models

The regression model described in the subsections above forces the assumption that the error variance is proportional to X_i^3 because $\text{Var}(Y_i) = \beta^3 X_i^3 / \lambda$. It seems intuitively obvious that other variance assumptions might be more reasonable. Let us consider the alternative model:

$$Y_i \sim \text{IG}(\beta X_i, \lambda_i), \qquad i = 1, 2, \ldots, n \tag{8.16}$$

with $\beta > 0$, $X_i > 0$, $\lambda_i > 0$. Instead of assuming that the error variance is proportional to X_i^3, we make the assumption that $\beta^2 X_i^2 / \lambda_i = k$. Under this assumption $Y_i \sim \text{IG}(\beta X_i, \beta^2 X_i^2 / k)$, $i = 1, 2, \ldots, n$, and a development can be made completely analogous to that in Section 8.2.1. This leads us to the estimate of β given by

$$\tilde{\beta} = \frac{\bar{Y}}{\bar{X}}. \tag{8.17}$$

This estimate has a strikingly familiar form and is often used in ratio estimation with many different models. Whitmore (1986a) considers the problem of estimating β when $Y_i \sim \text{IG}(\beta X_i^\gamma, k X_i^\rho)$ for given exponents γ and ρ. He notes that the transformed variables (Y_i', X_i'), where $Y_i' = X_i^{\rho - 2\gamma} Y_i$ and $X_i' = X_i^{\rho - \gamma}$, satisfy the model $Y_i' \sim \text{IG}(\beta X_i', k X_i'^2)$. This particular class of models includes both of the models already presented for the zero-intercept case.

Table 8.1 Projected and Actual Consumer-Product Sales ($000)

Product (i)	Projected (x_i)	Actual (y_i)	Product (i)	Projected (x_i)	Actual (y_i)
1	5,959	5,673	11	527	487
2	3,534	3,659	12	353	463
3	2,641	2,565	13	331	225
4	1,965	2,182	14	290	257
5	1,738	1,839	15	253	311
6	1,182	1,236	16	193	212
7	667	918	17	156	166
8	613	902	18	133	123
9	610	756	19	122	198
10	549	500	20	114	99

Source: Whitmore, 1986a.

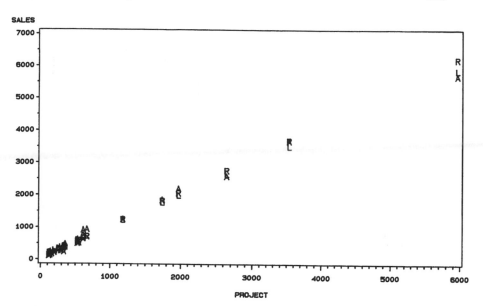

FIGURE 8.1 Comparison of least squares estimates with maximum likelihood estimates (Whitmore data). A, actual; R, ratio estimate; L, least square estimate.

Whitmore gives as an example the data in Table 8.1. The projected sales were made by a market survey organization.

The actual sales Y_i were assumed to be IG($\beta X_i, kX_i$) where X_i is the projected sales. A value of $\hat{\beta} = 1.0383$ was obtained. This particular set of data is so well behaved that almost any linear model may do very well. A simple linear regression model fitted by unweighted least squares gave $\hat{\alpha} = 76.7862$ and $\hat{\beta} = 0.9683$. The actual data as well as the ratio estimates and the least squares estimates are shown in Figure 8.1. It can be seen that the least squares estimates fit very well compared with the IG maximum likelihood ratio estimates.

8.3 SIMPLE LINEAR REGRESSION MODEL WITH INTERCEPT

Although there are several regression examples in the literature where the zero-intercept models just presented are reasonable, there is more interest in simple linear regression which does not assume a zero intercept. Unfortunately, such models have not allowed the sort of development given in the previous sections. However, working procedures are available because unbiased estimates of the parameters and variances of these estimates have been obtained.

Consider the model:

$$Y_i \sim \text{IG}(\alpha + \beta X_i, \lambda), \qquad \alpha > 0,\ \beta > 0,\ X_i > 0,\ \lambda > 0. \qquad (8.18)$$

Maximum likelihood has not yielded closed formulas for estimators. However, it can be verified that the following estimates are unbiased for β and α:

$$\hat{\beta} = \frac{\sum [(Y_i - \bar{Y})/X_i^2]}{\sum [(X_i - \bar{X})/X_i^2]}, \qquad (8.19)$$

$$\hat{\alpha} = \bar{Y} - \hat{\beta}\bar{X}. \qquad (8.20)$$

Next, if we assume Y_i is inverse Gaussian with mean $\alpha + \beta X_i$ and variance $(\alpha + \beta X_i)^3 \lambda_i^{-1}$ such that the ratio of the variance to the mean equals a constant k, we can also obtain the unbiased estimates $\hat{\alpha}$ and $\hat{\beta}$ where

$$\hat{\beta} = \frac{\sum [(X_i - \bar{X})(Y_i - \bar{Y})/X_i]}{\sum [(X_i - \bar{X})^2/X_i]},$$

$$\hat{\alpha} = \bar{Y} - \hat{\beta}\bar{X}. \qquad (8.21)$$

This follows because the additive property holds whenever this ratio is constant.

These estimators are suggested from the estimators in the previous section by correcting for the mean. Presumably, this idea would yield estimators for a more general class of intercept models similar to that considered by Whitmore (1986a) for the zero-intercept case.

8.4 NONLINEAR REGRESSION MODELS

8.4.1 Introduction

Except for the zero-intercept case, attempts to develop a linear model theory for the general model $y = x\beta + e$, where y is a vector of inverse Gaussian variables, have not proven fruitful. However, the work in Tweedie's papers suggests linear models may be possible for $1/Y_i$. Such models are, of course, nonlinear in the original variables. Also, it is possible to develop linear models for the reciprocal of the mean.

8.4.2 A Model with Restricted Parameter Space

Suppose we consider

$$Y_i \sim \text{IG}((\alpha + \beta X_i)^{-1}, \lambda), \qquad i = 1, 2, \ldots, n, \, X_i > 0, \tag{8.22}$$

where (α, β, λ) belongs to an appropriate parameter region R. Bhattacharyya and Fries (1982) take the region to be given by

$$R_1 = \{(\alpha, \beta, \lambda): \alpha \geqslant 0, \, \beta \geqslant 0, \, \alpha + \beta > 0, \, \lambda > 0\}. \tag{8.23}$$

The requirement that λ be positive is very natural. We might be interested in a different parameter space for α and β. Because the mean $(\alpha + \beta X)^{-1}$ must be positive and finite for an inverse Gaussian distribution, it may be reasonable to require that $\alpha + \beta X > 0$ for all X in some interval which contains all sample X values. We therefore suggest the region

$$R_2 = \{(\alpha, \beta, \lambda): \alpha + \beta X > 0 \text{ for } X_L < X < X_U, \, \lambda > 0\}. \tag{8.24}$$

The log-likelihood function is given by

$$L = \frac{n}{2} \log \lambda - \frac{\lambda}{2} \sum \frac{[(\alpha + \beta X_i)Y_i - 1]^2}{Y_i}$$

$$= \frac{n}{2} \log \lambda - \frac{\lambda}{2} [\alpha^2 n V_0 + 2\alpha\beta n V_1 + \beta^2 n V_2$$

$$- 2n\alpha - 2\beta \sum X_i + nR], \tag{8.25}$$

where, as given by Bhattacharyya and Fries,

$$nV_0 = \sum Y_i, \qquad nV_1 = \sum X_i Y_i, \qquad nV_2 = \sum X_i^2 Y_i,$$
$$nR = \sum (1/Y_i). \qquad (8.26)$$

Differentiating with respect to α, β, and λ gives the likelihood equations

$$\alpha n V_0 + \beta n V_1 - n = 0$$
$$\alpha n V_1 + \beta n V_2 - \sum X_i = 0$$
$$\frac{n}{2\lambda} - \frac{1}{2} \sum Y_i^{-1} [1 - (\alpha + \beta X_i) Y_i]^2 = 0. \qquad (8.27)$$

These lead to the roots of the log-likelihood equation

$$\alpha^* = \frac{1}{\bar{Y}} - \beta^* \frac{\sum X_i Y_i}{\sum Y_i} = \frac{1}{\bar{Y}} \left(1 - \beta \frac{\sum X_i Y_i}{n} \right)$$
$$= \frac{V_2 - \bar{X} V_1}{V_0 V_2 - V_1^2}$$
$$\beta^* = \frac{n \sum (X_i - \bar{X})(Y_i - \bar{Y})}{(\sum X_i Y_i)^2 - (\sum y_i)(\sum X_i^2 Y_i)}$$
$$= \frac{\bar{X} V_0 - V_1}{V_0 V_2 - V_1^2}$$
$$\lambda^{*-1} = \frac{1}{n} \sum Y_i^{-1} [1 - (\alpha^* + \beta^* X_i) Y_i]^2. \qquad (8.28)$$

From (8.26) it can be seen that λ^* will lie in R_1 or R_2 for any (α^*, β^*). However, it may be that for some sample (α^*, β^*) will lie outside the parameter space. From (8.27) we can see that at most one of α^* and β^* can be negative. To understand what to do if (α^*, β^*) is outside the parameter space, consider the likelihood function again. It can be shown that twice the negative of the exponent term can be written as the sum of two terms as follows:

$$\lambda \sum \frac{[(\alpha + \beta X_i)Y_i - 1]^2}{Y_i} = \frac{n\lambda}{\lambda^*} + \lambda \sum [(\alpha - \alpha^*)$$
$$+ (\beta - \beta^*)X_i]^2 Y_i. \qquad (8.29)$$

The left side in (8.29) is chi-square with n degrees of freedom. At one time it was conjectured that the two terms on the right were independent chi-squares with $n - 2$ and two degrees of freedom, respectively. However, Letac and Seshadri (1986) showed this to be false. From (8.29) it can be seen that for any λ^*, the contours for α and β are concentric ellipses centered on (α^*, β^*). Thus, if (α^*, β^*) lies in the second or fourth quadrant (β on the vertical and α on the horizontal axis), the likelihood is maximized by searching on the boundaries of R_1 or R_2. Suppose $\alpha^* < 0$. Then the maximum likelihood for R_1 is achieved when $\alpha = 0$. Setting $\alpha = 0$ in (8.25) and differentiating with respect to β gives

$$\beta n V_2 - \sum X_i = 0. \qquad (8.30)$$

Proceeding in this way, we get the results obtained by Bhatta-charyya and Fries for parameter region R_1:

$$(\hat{\alpha}, \hat{\beta}) = (\alpha^*, \beta^*) \qquad \text{if } V_1 < \min(\bar{X} V_0, \bar{X}^{-1} V_2)$$
$$= (0, \bar{X} V_2^{-1}) \qquad \text{if } \bar{X}^{-1} V_2 \leqslant V_1 \leqslant \bar{X} V_0$$
$$= (V_0^{-1}, 0) \qquad \text{if } \bar{X} V_0 \leqslant V_1 \leqslant \bar{X}^{-1} V_2,$$
$$\hat{\lambda}^{-1} = R - \hat{\alpha} - \hat{\beta} \bar{X}. \qquad (8.31)$$

Now suppose that the parameter space R_2 is considered. Again if $\alpha^* < 0$ and (α^*, β^*) is below the line $\alpha + \beta X_L = 0$, then the maximum likelihood is maximized by letting $\alpha = -\beta X_L$ and then maximizing L with respect to β. Proceeding in this way yields the maximum likelihood estimate for parameter region R_2:

$$(\hat{\alpha}, \hat{\beta}) = (\alpha^*, \beta^*) \qquad \text{if } (\alpha^*, \beta^*) \in R_2$$

$$= \left(-\hat{\beta}X_L, \hat{\beta} = \frac{\sum X_i - nX_L}{n(X_L^2 V_0 + V_2 - 2V_1 X_L)} \right)$$

$$\text{if } \alpha^* < 0 \text{ and } (\alpha^*, \beta^*) \text{ is below } \alpha + \beta X_L = 0$$

$$= \left(-\hat{\beta}X_U, \hat{\beta} = \frac{\sum X_i - nX_U}{n(X_U^2 V_0 + V_2 - 2V_1 X_U)} \right)$$

$$\text{if } \beta^* < 0 \text{ and } (\alpha^*, \beta^*) \text{ is below } \alpha + \beta X_U = 0,$$

$$\tilde{\lambda}^{-1} = R - \tilde{\alpha} - \tilde{\beta}\bar{X}. \tag{8.32}$$

Bhattacharyya and Fries use asymptotic theory to give approximate variances for the MLEs as follows:

$$\text{Var}(\hat{\alpha}) \doteq (n\hat{\lambda}D)^{-1}V_2$$
$$\text{Var}(\hat{\beta}) \doteq (n\hat{\lambda}D)^{-1}V_0$$

and

$$\text{Var}(\hat{\lambda}) \doteq 2n^{-1}\hat{\lambda}^2. \tag{8.33}$$

These can be used to establish confidence intervals. Presumably, the same asymptotic variables hold for the root estimators as well as the MLE estimators for R_2. Bhattacharyya and Fries (1982) use the data reported by Nelson (1971) for failure times of an insulation material in a motorette test performed at four temperature settings ranging from 190 to 260°C. The data are given in Table 8.2. Excluding the 260°C data (see Nelson and Bhattacharyya and Fries for discussion) and letting $X_i = 10^{-8}[T^3 - 180^3]$ and $Y_i = $ (hours to failure)/1000 we obtain

$$\bar{X} = 0.04611667$$
$$\bar{Y} = 4.33373333$$
$$\quad = V_0$$

Table 8.2 Hours to Failure

190°C	220°C	240°C	260°C
7228	1764	1175	600
7228	2436	1175	744
7228	2436	1521	744
8448	2436	1569	744
9167	2436	1617	912
9167	2436	1665	1128
9167	3108	1665	1320
9167	3108	1713	1464
10511	3108	1761	1608
10511	3108	1953	1896

Source: Nelson, 1971.

$$R = \frac{1}{n}\sum Y_i^{-1}$$

$$= 0.38489068$$
$$V_1 = 0.11453517$$
$$V_2 = 0.00571487. \tag{8.34}$$

Using these values we obtain

$$\alpha^* = 0.0371633$$
$$\beta^* = 7.32477$$
$$\lambda^{*-1} = 0.00993324. \tag{8.35}$$

These results are in close agreement with those reported by Bhattacharyya and Fries.

Despite the inherent beauty of the preceding development, a nonlinear model carries some disadvantages. Interpretation of the parameters becomes difficult. We are tempted to call β the slope but it is not the slope, and so on. A special property of the inverse Gaussian distribution is that the mean and the variance of the reciprocal of an inverse Gaussian variable are

$$E\left(\frac{1}{Y}\right) = \frac{1}{\mu} + \frac{1}{\lambda}$$

$$\text{Var}\left(\frac{1}{Y}\right) = \frac{1}{\mu\lambda} + \frac{2}{\lambda^2}. \tag{8.36}$$

In the present setting then

$$E\left(\frac{1}{Y_i}\right) = \alpha + \beta X_i + \frac{1}{\lambda}$$

$$\text{Var}\left(\frac{1}{Y_i}\right) = \frac{\alpha + \beta X_i}{\lambda} + \frac{2}{\lambda^2}. \tag{8.37}$$

This suggests regressing the reciprocal of Y on X using a simple linear regression model. Bhattacharyya and Fries correct the reciprocals of Y for the bias $1/\lambda$ and then regress the bias corrected reciprocals on X using weighted least squares. We also fitted the present data by unweighted least squares, which gave a poorer fit.

This data set does not enable us to assess the goodness of the regression model very well because there are so few X values. Further, there appears to be a grouping error in the reported lifetimes.

8.4.3   A General Nonlinear Regression Model

Consider generalizing the model in (8.22). Suppose that

$$Y_i \sim \text{IG}\left(\frac{1}{\mathbf{x}_i\boldsymbol{\beta}}, \lambda\right), \qquad i = 1, \dots, n, \tag{8.38}$$

where $\mathbf{x}_i$ is a $1 \times p$ row vector and $\boldsymbol{\beta}$ is a $p \times 1$ column vector. Then the joint density function for $Y_1, Y_2, \dots, Y_n$ can be written as

$$\left(\frac{\lambda}{2\pi}\right)^{n/2} \prod_1^n Y_i^{-3/2} \exp - \frac{\lambda}{2} (\mathbf{WX}\boldsymbol{\beta} - \mathbf{J})'\mathbf{W}^{-1}(\mathbf{WX}\boldsymbol{\beta} - \mathbf{J}),$$

$$\tag{8.39}$$

where $\mathbf{W}$ is an $n \times n$ diagonal matrix with $Y_1, Y_2, Y_3, \ldots, Y_n$ on the diagonal and $\mathbf{J}$ is an $n \times 1$ vector of ones.

Differentiating with respect to β and λ gives the striking results

$$\hat{\beta} = (\mathbf{X'WX})^{-1}\mathbf{X'J}$$

$$\frac{1}{\hat{\lambda}} = \frac{1}{n}(\mathbf{J'W}^{-1}\mathbf{J} - \mathbf{J'X\hat{\beta}}). \qquad (8.40)$$

These formulas are remarkably similar to the estimates of β and σ^2 with normal distribution theory. Further, minus twice the term in the exponent in (8.39) decomposes into

$$\lambda(\hat{\beta} - \beta)'\mathbf{X'WX}(\hat{\beta} - \beta) + \frac{n\lambda}{\hat{\lambda}}. \qquad (8.41)$$

Whitmore once conjectured that the two terms in (8.41) were independent chi-squared random variables with p and $n - p$ degrees of freedom, respectively. However Letac and Seshadri (1986) showed this conjecture to be false by presenting a counterexample.

Whitmore (1983) gives asymptotic theory for distribution of the maximum likelihood estimates in (8.40). He applies his results to failure data when the data are censored.

8.4.4   A Model with Restricted Intercept

When $Y \sim IG(\mu, \lambda)$, the reciprocal $W = 1/Y = U + V$ where $U \sim IG(1/\mu, \lambda/\mu^2)$ and $V \sim (1/\lambda)\chi_1^2$ are independent. Also, $E(W) = E(U) + E(V) = 1/\mu + 1/\lambda$. This suggests that one may consider a linear model for W as

$$W = \alpha + \beta X + \varepsilon, \qquad (8.42)$$

where $\beta X = E(U)$ and $\alpha = E(V)$, $X > 0$. Clearly $\varepsilon = W - \alpha - \beta X$. Therefore the distribution of ε is a three-parameter distribution corresponding to that of the reciprocal of an inverse Gaussian variable with a threshold parameter of $-\alpha - \beta X$.

Estimation of β and α can be made by considering $Y_i \sim IG((\beta X_i)^{-1}, \alpha^{-1})$, $i = 1, 2, \ldots, n$, independent random variables. From the discussion given in Section 8.2, it follows that the maximum likelihood estimates are

$$\hat{\beta} = \frac{\sum X_i}{\sum X_i^2 Y_i}$$

$$\hat{\alpha} = \bar{W} - \hat{\beta}\bar{X}, \tag{8.43}$$

where

$$\bar{W} = \frac{1}{n}\sum\frac{1}{Y_i}.$$

$1/\hat{\beta}$ is a linear combination of Y_i's and has an inverse Gaussian distribution. It can be easily shown that

$$\sum X_i^2 Y_i \sim \text{IG}(\beta^{-1}\sum X_i, \alpha^{-1}(\sum X_i)^2). \tag{8.44}$$

So

$$\frac{1}{\hat{\beta}} \sim \text{IG}\left(\frac{1}{\beta}, \alpha^{-1}(\sum X_i)\right). \tag{8.45}$$

Thus $1/\hat{\beta}$ is an unbiased estimate of $1/\beta$. On the other hand, we have $E[\hat{\beta}] = \beta + \alpha/\sum X_i$ and $\hat{\beta}$ given in (8.43) is a biased estimate of β. Since $E[\bar{W}] = \alpha + \beta\bar{X}$, it follows that $E[\hat{\alpha}] = [(n-1)/n]\alpha$. Accordingly $[n/(n-1)]\hat{\alpha}$ is an unbiased estimate of α.

Next, $\hat{\beta}$ and $\hat{\alpha}$ are independently distributed. One may see this by recognizing $1/\beta$ (up to a constant) and $1/\alpha$ in the role of standard inverse Gaussian parameters μ and λ of $\text{IG}(\mu, \lambda)$. Thus all inferential properties of α^{-1} and β^{-1} can simply be established by reworking the results given earlier in terms of these parameters.

9

Life Testing and Reliability

9.1 INTRODUCTION

As stated in Chapter 1, we consider the inverse Gaussian distribution very relevant for studying reliability and life-testing problems. The inverse Gaussian being the first passage time distribution for the Wiener process makes it particularly appropriate for failure or reaction time data analysis.

Suppose F denotes the distribution function of failure time for a mechanism. Then the reliability $R(x)$ of the mechanism at time x is defined by the probability of its having no failure before time x; thereby, $R(x) = 1 - F(x)$. It now follows from equation (2.14) that the reliability function for the inverse Gaussian is

$$R(x) = \Phi\left(\sqrt{\frac{\lambda}{x}}\left(1 - \frac{x}{\mu}\right)\right) - e^{2\lambda/\mu}\Phi\left(-\sqrt{\frac{\lambda}{x}}\left(1 + \frac{x}{\mu}\right)\right). \tag{9.1}$$

The reliability function for several values of λ, when $\mu = 1$, is depicted in Figure 9.1.

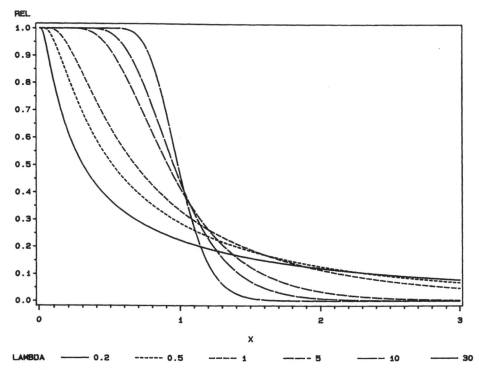

Figure 9.1   Reliability function for $\mu = 1$ for six values of λ.

In this chapter we first consider estimation of $R(x)$ given in (9.1). The minimum variance unbiased estimate (MVUE) of $R(x)$ is derived in the next section. Following that, we discuss other aspects of reliability such as the failure rate and the mean residual life of the inverse Gaussian. Certain tolerance and prediction limits are obtained. Last, comparisons between the inverse Gaussian and some other distributions used in reliability studies are discussed.

9.2   ESTIMATION OF RELIABILITY FUNCTION

9.2.1   Introduction

The MLE of the reliability function $R(x)$ is obtained by replacing μ and λ in (9.1) by their estimates $\hat{\mu}$ and $\hat{\lambda}$ given in (5.2). Accordingly, the MLE of $R(x)$ is

$$\hat{R}(x) = \Phi\left[\sqrt{\frac{\hat{\lambda}}{x}}\left(1 - \frac{x}{\hat{\mu}}\right)\right] - e^{2\hat{\lambda}/\hat{\mu}}\Phi\left[-\sqrt{\frac{\hat{\lambda}}{x}}\left(1 + \frac{x}{\hat{\mu}}\right)\right], \qquad (9.2)$$

where

$$\hat{\mu} = \bar{X} = \frac{1}{n}\sum_{i=1}^{n} X_i \qquad \text{and} \qquad \hat{\lambda}^{-1} = \frac{1}{n}\sum_{i=1}^{n}\left(\frac{1}{X_i} - \frac{1}{\bar{X}}\right)$$

for a random sample $X_1, X_2, \ldots, X_n$ from $IG(\mu, \lambda)$.

Chhikara and Folks (1974) gave the MVUEs of the inverse Gaussian cdf, $F(x)$, for different cases of unknown parameters. These estimates were shown to have a close analogy with their counterparts for the normal distribution function. The MVUE of $R(x)$ when both μ and λ are unknown, obtained directly from that of $F(x)$, was compared with the MLE given in (9.2) for a set of repair times data in Chhikara and Folks (1977).

The minimum variance unbiased estimation approach is to find any unbiased estimate and then to obtain its expected value conditional on a given complete sufficient statistic. Let $X = (X_1, X_2, \ldots, x_n)$ be a sample from an inverse Gaussian population with parameter $\theta = (\mu, \lambda)$. Then $I_{[x,\infty)}(X_1)$ is an unbiased estimate of $R(x; \theta)$. If $T(X)$ is a complete sufficient statistic, the MVUE of $R(x; \theta)$ is given by $E[I_{[x,\infty)}(X_1)|T(X)]$, and it will be denoted by $\hat{R}(x; \theta)$. Here $I_{[\cdot]}$ denotes the indicator function. We will consider the problem of estimation for all three cases: (1) μ unknown and λ known, (2) μ known and λ unknown, and (3) both μ and λ unknown. Our results will be given in the forms which can be evaluated by using the standard normal table for case 1 and by using Student's t table for cases 2 and 3.

9.2.2 MVUE of $R(x; \theta)$ When λ Is Known

The sample mean $\bar{X} = \sum_1^n X_i/n$ is a complete sufficient statistic and to derive the MVUE of $R(x; \mu)$, we first find the conditional density function of X_1, given $\bar{X} = \bar{x}$.

The joint density function of X_1 and $\bar{X}$ is obtained as

$$f(x_1, \bar{x}) = \frac{n(n-1)\lambda}{2\pi[x_1(n\bar{x} - x_1)]^{3/2}}$$

$$\times \exp\left\{-\frac{\lambda}{2\pi^2}\left[\frac{(x_1 - \mu)^2}{x_1} + \frac{[n(\bar{x} - \mu) - (x_1 - \mu)]^2}{n\bar{x} - x_1}\right]\right\},$$

$$0 < x_1 < n\bar{x}.$$

Since $\bar{X} \sim IG(\mu, n\lambda)$, it follows that the conditional density function $h(x_1 \mid \bar{x})$ of X_1, given $\bar{X} = \bar{x}$, is given by the ratio $f(x_1, \bar{x})/g(\bar{x})$, where $g(\bar{x})$ is the density function of $\bar{X}$, and this ratio simplifies to

$$h(x_1 \mid \bar{x}) = \sqrt{\frac{n\lambda}{2\pi}} \frac{(n-1)\bar{x}^{3/2}}{[x_1(n\bar{x} - x_1)]^{3/2}} \exp\left[-\frac{n\lambda(x_1 - \bar{x})^2}{2x_1\bar{x}(n\bar{x} - x_1)}\right], \quad (9.3)$$

$$0 < x_1 < n\bar{x}.$$

Hence the MVUE of $R(x; \mu)$ is given by

$$\hat{R}(x; \mu) = \int_x^{n\bar{x}} h(x_1 \mid \bar{x}) \, dx_1, \qquad x > 0 \qquad\qquad (9.4)$$

with $h(x_1 \mid \bar{x})$ as in (9.3).

Next we express the right side of (9.4) in the form in which the standard normal table can be used to evaluate $\hat{R}(x; \mu)$ for all $x > 0$. Let $w = \sqrt{n\lambda}(x_1 - \bar{x})/\sqrt{x_1\bar{x}(n\bar{x} - x_1)}$. We have a one-to-one transformation, and w varies from $-\infty$ to ∞ as x_1 varies from 0 to $n\bar{x}$. Then, using the approach given in Chhikara and Folks (1974) in estimating $F(x)$ to simplify the right side in (9.4), we obtain

$$\hat{R}(x; \mu) = \frac{1}{\sqrt{2\pi}} \int_{w'}^{\infty} \left[1 - \frac{(n-2)w\sqrt{\bar{x}}}{\sqrt{4n(n-1)\lambda + n^2\bar{x}w^2}}\right] e^{-w^2/2} \, dw,$$

where

$$\qquad\qquad\qquad\qquad\qquad\qquad\qquad\qquad\qquad\qquad\qquad\qquad (9.5)$$

$$w' = \frac{\sqrt{n\lambda}(x - \bar{x})}{\sqrt{x\bar{x}(n\bar{x} - x)}}.$$

The integral in (9.5) can be expressed in terms of the standard normal distribution for a given value of w'. Separating the integrand and letting $u = [w^2 + 4(n - 1)\lambda/n\bar{x}]^{1/2}$, it can be shown that the second term on the right side simplifies to

$$\frac{n-2}{n} e^{2(n-1)\lambda/n\bar{x}} \Phi\left(-\sqrt{\frac{4(n-1)\lambda}{n\bar{x}} + w'^2}\right)$$

Accordingly, (9.5) reduces to

$$\hat{R}(x; \mu) = \Phi(-w') - \frac{n-2}{n} e^{2(n-1)\lambda/n\bar{x}} \Phi\left(-\sqrt{\frac{4(n-1)\lambda}{n\bar{x}} + w'^2}\right),$$

$$-\infty < w' < \infty.$$

Hence the MVUE of $R(x; \mu)$ is

$$\hat{R}(x; \mu) = \begin{cases} 0, & x > n\bar{x} \\ 1, & x < 0 \\ \Phi(-w') - \dfrac{n-2}{n} e^{2(n-1)\lambda/n\bar{x}} \Phi(-w''), & \text{otherwise,} \end{cases}$$

where $\qquad$ (9.6)

$$w' = \frac{\sqrt{n\lambda}(x - \bar{x})}{\sqrt{x\bar{x}(n\bar{x} - x)}} \quad \text{and} \quad w'' = \frac{\sqrt{\lambda}(n\bar{x} + (n - 2)x)}{\sqrt{n\bar{x}x(n\bar{x} - x)}}.$$

9.2.3 MVUE of $R(x; \theta)$ When μ Is Known

The statistic $T(X) = \sum_1^n (X_i - \mu)^2/X_i$ is complete sufficient and $\lambda T(X)/\mu^2$ has the χ^2 distribution with n degrees of freedom. Following Chhikara and Folks (1974), the conditional density function of X_1 given $T(X) = t$ is

$$h(x_1 \mid t) = \frac{\mu}{B(\frac{1}{2}, (n - 1)/2)} \frac{1}{(tx_1^3)^{1/2}} \left[1 - \frac{(x_1 - \mu)^2}{tx_1}\right]^{(n-3)/2},$$

$$0 < \frac{(x_1 - \mu)^2}{x_1} < t. \qquad (9.7)$$

We can now obtain the MVUE of $R(x; \lambda)$ given by

$$\hat{R}(x; \lambda) = \int_x^U h(x_1 \,|\, t)\, dx, \qquad x > L \tag{9.8}$$

where

$$L = \tfrac{1}{2}(2\mu + t - \sqrt{4\mu t + t^2})$$

and

$$U = \tfrac{1}{2}(2\mu + t + \sqrt{4\mu t + t^2}) \tag{9.9}$$

with $h(x_1 \,|\, t)$ as in (9.7).

Next we simplify (9.8) so that $\hat{R}(x; \mu)$ can be evaluated using Student's t percentage points. Let

$$w = \frac{\sqrt{n-1}(x_1 - \mu)}{\sqrt{tx_1}}\left[1 - \frac{(x_1 - \mu)^2}{tx_1}\right]^{-1/2}.$$

This is a one-to-one transformation and $-1 < (x_1 - \mu)/\sqrt{tx_1} < 1$ implies $w \in (-\infty, \infty)$. Next

$$\frac{dx_1}{dw} = \left[\frac{dw}{dx_1}\right]^{-1} = \frac{2(tx_1^3)^{1/2}}{\sqrt{n-1}(x_1 + \mu)}\left[1 + \frac{(x_1 - \mu)^2}{tx_1}\right]^{3/2} \tag{9.10}$$

and

$$x_1 =$$

$$\frac{\left[2\mu\left(1 + \dfrac{w^2}{n-1}\right) + t\,\dfrac{w^2}{n-1}\right] + \dfrac{w}{\sqrt{n-1}}\left[4\mu t\left(1 + \dfrac{w^2}{n-1}\right) + t^2\,\dfrac{w^2}{n-1}\right]^{1/2}}{2\left(1 + \dfrac{w^2}{n-1}\right)}.$$

$$\tag{9.11}$$

since $w \in (-\infty, 0) \Leftrightarrow x_1 \in (L, \mu)$ and $w \in (0, \infty) \Leftrightarrow x_1 \in (\mu, U)$ where L and U are given in (9.9). By substituting in (9.8) from (9.7), (9.10), and (9.11) and then simplifying we have

$$\hat{R}(x; \lambda) = G_{t,n-1}(w') - \left(1 + \frac{4\mu}{t}\right)^{n/2-1}$$

$$\times G_{t,n-1}\left(\sqrt{n-1}\left[\frac{4\mu}{t}\left(1 + \frac{w'^2}{n-1}\right) + \frac{w'^2}{n-1}\right]^{1/2}\right),$$

$$-\infty < w' < \infty$$

where $G_{t,n-1}$ denotes the right tail area under the curve of the Student's t distribution with $(n-1)$ df and w' is as given below in (9.12). Accordingly, the MVUE of $R(x; \lambda)$ is

$$\hat{R}(x; \lambda) = \begin{cases} 0, & x > \frac{1}{2}[(2\mu + t) + \sqrt{4\mu t + t^2}] \\ 1, & x < \frac{1}{2}[(2\mu + t) - \sqrt{4\mu t + t^2}] \\ G_{t,n-1}(w') - \left[\dfrac{t + 4\mu}{t}\right]^{(n/2-1)} G_{t,n-1}(w''), & \text{otherwise} \end{cases}$$

$$(9.12)$$

where

$$w' = \frac{\sqrt{n-1}(x-\mu)}{\sqrt{tx - (x-\mu)^2}}, \qquad w'' = \frac{\sqrt{n-1}(x+\mu)}{\sqrt{tx - (x-\mu)^2}}$$

and $t = \sum (x_i - \mu)^2 / x_i$.

9.2.4   MVUE of $R(x; \theta)$ When Both μ and λ Are Unknown

The statistic $T(X) = (\bar{X}, V)$, where $\bar{X} = (1/n)\sum X_i$ and $V = \sum (1/X_i - 1/\bar{X})$, forms a complete sufficient statistic. The density function of X_1 given $\bar{X} = \bar{x}$ and $V = v$ is given in (5.8). Accordingly, the MVUE of $R(x; \mu, \lambda)$ is

$$\hat{R}(x; \mu, L) = \int_x^U h(x_1 \mid T(x)) \, dx_1 \qquad (9.13)$$

with $h(x_1 \mid T(x))$ as given in (5.8).

Next the integral in (9.13) can be simplified so that $\hat{R}(x; \mu, \lambda)$ can be evaluated using percentile values from a Student's t table. The procedure is the same as used in the previous two cases. One may refer to Chhikara and Folks (1974) for specific details. When the right side of (9.13) is simplified, we have

$$\hat{R}(x; \mu, \lambda) = \frac{1}{B(\frac{1}{2},(n-2)/2)}$$

$$\times \int_{w'}^{\infty} \left[1 - \frac{(n-2)w\sqrt{v\bar{x}}}{\sqrt{4n(n-1)(1+w^2) + n^2 v\bar{x}w^2}} \right]$$

$$\times [1 + w^2]^{-(n-1)/2} \, dw \tag{9.14}$$

where

$$w' = \frac{\sqrt{n}(x - \bar{x})}{\sqrt{vx\bar{x}(n\bar{x} - x) - n(x - \bar{x})^2}} \, .$$

Hence it follows that the MVUE of $R(x; \mu, \lambda)$ is

$$\hat{R}(x; \mu, \lambda) =$$

$$\begin{cases} 0, & x > U \\ 1, & x < L \\ G_{t,n-2}(w'') - \dfrac{n-2}{n} \left[1 + \dfrac{4(n-1)}{nv\bar{x}} \right]^{(n-3)/2} G_{t,n-2}(w'''), & \text{otherwise,} \end{cases}$$

$$\tag{9.15}$$

where $G_{t,n-2}$ denotes the right tail area under the curve of Student's t distribution with $(n-2)$ df,

$$w'' = \frac{\sqrt{n(n-2)}(x - \bar{x})}{\sqrt{vx\bar{x}(n\bar{x} - x) - n(x - \bar{x})^2}},$$

$$w''' = \frac{\sqrt{n(n-2)}[\bar{x} + (n-2)x/n]}{\sqrt{vx\bar{x}(n\bar{x} - x) - n(x - \bar{x})^2}}, \tag{9.16}$$

and

$$L = \frac{\bar{x}}{2(n + v\bar{x})} \left[n(2 + v\bar{x}) - \sqrt{4n(n - 1)v\bar{x} + n^2 v^2 \bar{x}^2} \right]$$

$$U = \frac{\bar{x}}{2(n + v\bar{x})} \left[n(2 + v\bar{x}) + \sqrt{4n(n - 1)v\bar{x} + n^2 v^2 \bar{x}^2} \right]. \qquad (9.17)$$

Comparison of the estimates given in (9.6), (9.12), and (9.15) with those of their counterparts for the normal distribution (see Folks, Pierce, and Stewart, 1965) shows a remarkable similarity. Not only are their MVUEs commonly expressed in terms of the same distributions, standard normal, and Student's t, but these estimates are also similar in form and with respect to the parameters involved. The obvious difference that the MVUEs obtained for the inverse Gaussian are expressed as the nonlinear combination of standard normal distributions or of Student's t distributions is expected since the inverse Gaussian distribution is itself a nonlinear weighted normal distribution as discussed in Section 2.5.

Example

The following maintenance data were reported on active repair times (hours) for an airborne communication transceiver (Von Alven, 1964, p. 156):

.2, .3, .5, .5, .5, .6, .6, .7, .7, .7, .8, .8, 1.0, 1.0, 1.0, 1.0, 1.1, 1.3, 1.5, 1.5, 1.5, 1.5, 2.0, 2.0, 2.2, 2.5, 2.7, 3.0, 3.0, 3.3, 3.3, 4.0, 4.0, 4.5, 4.7, 5.0, 5.4, 5.4, 7.0, 7.5, 8.8, 9.0, 10.3, 22.0, 24.5.

Chhikara and Folks (1977) considered the inverse Gaussian model for these repair times and showed that it provides a good fit to the data. Von Alven fitted the lognormal model. The observed values of the Kolmogorov-Smirnov test statistic are .053 for the inverse Gaussian and .081 for the lognormal, indicating both models provide equally good fits.

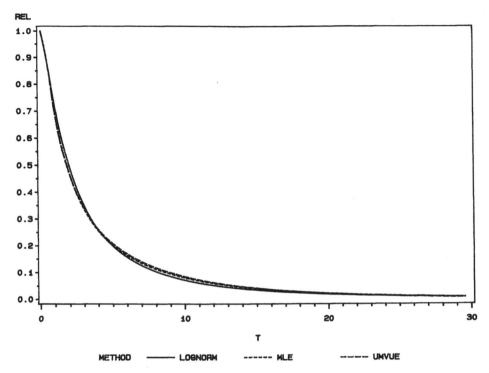

Figure 9.2 Comparison of estimators of $R(t)$—Von Alven data.

The MVUEs μ and $1/\lambda$ are given by

$$\bar{x} = 3.61 \quad \text{and} \quad \hat{\lambda}^{-1} = \frac{1}{n-1} \sum_1^n (x_i^{-1} - \bar{x}^{-1}) = .60,$$

respectively.

The MLE and MVUE of $R(x)$, $x > 0$, obtained from (9.2) and (9.15) are plotted in Figure 9.2 as well as the MLE of $R(x)$ for the lognormal model. The two estimates for the inverse Gaussian are almost the same. The MLE for the lognormal is slightly closer to the MLE for the inverse Gaussian.

9.3 TOLERANCE LIMITS

We consider the problem of finding a lower tolerance limit $L(X)$ based on a random sample $X = (X_1, X_2, \ldots, X_n)$ from IG(μ, λ) such that $\mathrm{Prob}\{R[L(X); \theta] > p\} = \gamma$ where

$$R[L(X); \theta] = \int_{L(X)}^{\infty} f(x; \theta) \, dx$$

and θ denotes the set of unknown parameters, for a specified tolerance level p and confidence level γ. Three cases arise regarding θ: (i) $\theta = \mu$, assuming λ known, (iii) $\theta = \lambda$, assuming μ known, and (iii) $\theta = (\mu, \lambda)$.

In general, suppose $\theta^*(X)$ is a 100γ percent lower confidence limit for θ, that is, $\mathrm{Prob}[\theta^*(X) < \theta] = \gamma$, and $L(X)$ is considered so that $R[L(X); \theta^*(X)] = p$. If $R(x; \theta)$ is a monotone increasing function in θ, one can easily see that $R[L(X); \theta] > p$ if and only if $\theta > \theta^*(X)$, implying $\mathrm{Prob}\{R[L(X); \theta] > p\} = \gamma$. Thus a lower tolerance limit is obtained by solving the equation

$$R[L(X); \theta^*(X)] = p \tag{9.18}$$

for $L(X)$.

Case i: μ Unknown and λ Known

For $X \sim \mathrm{IG}(\mu, \lambda)$, $R(x; \mu)$ is a monotone nondecreasing function in μ for any fixed λ and x as proved in Chhikara (1972). It follows from the UMP test given in (6.6) that a 100γ percent lower confidence bound for μ is

$$\mu^* = \bar{X}\left[1 + C\left(\frac{\bar{X}}{n\lambda}\right)^{1/2}\right]^{-1}, \tag{9.19}$$

where C is the solution of equation (6.7) with $(1 - \alpha)$ replaced by γ. For a determination of C, use the approximation discussed previously; that is, take $C = Z_\gamma$, the γ percentage point of the standard normal distribution. Now due to (9.18) a lower tolerance limit $L(X)$ is given by

$$p = \Phi\left[-\sqrt{\frac{\lambda}{L(\bar{X})}}\left(\frac{L(X)}{\mu^*} - 1\right)\right] - e^{2\lambda/\mu^*}\Phi\left[-\sqrt{\frac{\lambda}{L(X)}}\left(\frac{L(X)}{\mu^*} + 1\right)\right], \tag{9.20}$$

where Φ denotes the standard normal cdf. For any given λ and μ^*, $L(X)$ can be easily obtained from (9.20). (We assume that γ and p are already specified). However, $L(X)$ cannot be expressed explicitly in terms of λ, μ^* and percentage points of the standard normal distribution. The second term on the right side of (9.20) is equal to or less than $\Delta = \exp(2\lambda/\mu^*)\Phi(-2\sqrt{\lambda/\mu^*})$. Thus

$$
\Phi\left[-\sqrt{\frac{\lambda}{L(X)}}\left(\frac{L(X)}{\mu^*}-1\right)\right] - \Delta
$$

$$
\leqslant p \leqslant \Phi\left[-\sqrt{\frac{\lambda}{L(X)}}\left(\frac{L(X)}{\mu^*}-1\right)\right].
$$

Consequently, it can be shown that $L(X)$ is bounded by

$$
\mu^* + \left(\frac{\mu^{*2}}{2\lambda}\right)\left(Z_{p+\Delta}^2 + Z_{p+\Delta}\sqrt{Z_{p+\Delta}^2 + \frac{4\lambda}{\mu^*}}\right) \leqslant L(X)
$$

$$
\leqslant \mu^* + \left(\frac{\mu^{*2}}{2\lambda}\right)\left(Z_p^2 + Z_p\sqrt{Z_p^2 + \frac{4\lambda}{\mu^*}}\right), \tag{9.21}
$$

where Z_p and $Z_{p+\Delta}$ are the $(1 - p)$ and $(1 - p - \Delta)$ percentage points of the standard normal distribution.

When λ is large, Δ is small and can be neglected. Then

$$
L(X) = \mu^* + \left(\frac{\mu^{*2}}{2\lambda}\right)\left(Z_p^2 + Z_p\sqrt{Z_p^2 + \frac{4\lambda}{\mu^*}}\right). \tag{9.22}
$$

Case ii: μ Known and λ Unknown

In the case where μ is known and λ is unknown the monotone property does not hold for $R(x; \lambda)$. For $x \leqslant \mu$, $R(x; \lambda)$ is a nondecreasing function in λ for any fixed μ and x. If $x > \mu$, $R(x, \lambda)$ first increases and then decreases as λ increases. See Padgett (1979) for details. Thus the general method of obtaining a lower tolerance limit discussed above does not apply. Instead, one may proceed as follows:

Let λ_0 be the point of maxima of $R(x; \lambda)$ for $x > \mu$. Then, if

$x > \mu$, $R(x; \lambda)$ is nondecreasing for $\lambda \leqslant \lambda_0$ and is decreasing for $\lambda > \lambda_0$. It is easily seen from Section 6.2.3 that a 100γ percent lower confidence bound for λ is

$$\lambda^* = \frac{\mu^2 T}{\chi_{n,1-\gamma}^2},$$

(9.23)

where $T = \sum (X_i - \mu)^2 / X_i$ and $\chi_{n,1-\gamma}^2$ is the $100(1 - \gamma)$ percentage point of a chi-square distribution with n degrees of freedom. Thus it follows from (9.18) that a lower tolerance limit, $L(X)$, is given by the solution of the following equation:

$$p = \Phi\left[-\sqrt{\frac{\lambda^*}{L(X)}} \left(\frac{L(X)}{\mu} - 1 \right) \right]$$

$$- e^{2\lambda^*/\mu} \Phi\left[-\sqrt{\frac{\lambda^*}{L(X)}} \left(\frac{L(X)}{\mu} + 1 \right) \right]$$

(9.24)

except when $L(X) > \mu$ and $\lambda > \lambda_0$, in which case a lower tolerance limit is obtained as in (9.24) with λ^* replaced by $\tilde{\lambda}$, where

$$\tilde{\lambda} = \frac{\mu^2 T}{\chi_{n,\gamma}^2}$$

(9.25)

is obtained using the 100γ percentage point of chi-square distribution with n degrees of freedom. Since λ is unknown, one may use its estimate $\tilde{\lambda}$ to check against λ_0, which is a function of x and μ only. Padgett (1979) has used this approach to obtain confidence bounds for $R(x; \lambda)$ and showed that it provides fairly reliable lower confidence bounds.

Next one can obtain bounds for $L(X)$ similar to those given in case i. It follows that $L(X)$ is bounded by

$$\mu + \left(\frac{\mu^2}{2\lambda^*} \right)\left(Z_{p+\Delta}^2 + Z_{p+\Delta}\sqrt{Z_{p+\Delta}^2 + \frac{4\lambda^*}{\mu}} \right) \leqslant L(X)$$

$$\leqslant \mu + \left(\frac{\mu^2}{2\lambda^*} \right)\left(Z_p^2 + Z_p\sqrt{Z_p^2 + \frac{4\lambda^*}{\mu}} \right),$$

(9.26)

where Δ is equal to $\exp(2\lambda^*/\mu)\Phi(-2\sqrt{\lambda^*/\mu})$, except when $\tilde{\lambda} > \lambda_0$ and $L(X) > \mu$, in which case replace λ^* by $\tilde{\lambda}$.

Case iii: Both μ and λ Unknown

The preceding method of finding a lower tolerance limit can be applied provided there exists a parametric function, say $g(\theta)$, such that $R(x;\theta)$ is a monotone increasing or decreasing function in $g(\theta)$. At present we know of no such function and hence no solution to the problem is offered.

9.4   PREDICTION LIMITS

Often in reliability work one may want to construct a prediction interval for future random observations. Suppose we are given a random sample $\mathbf{X} = (X_1, X_2, \ldots, X_n)$ and want to have 100β percent confidence limits for an additional observation, say Y. Then the problem is equivalent to determining an interval $J(\mathbf{X})$ such that $J(\mathbf{X})$ covers on the average a β proportion of the distribution of Y, that is, $J(\mathbf{X})$ is the function of observations X_i's satisfying

$$E_X\left[\int_{J(\mathbf{X})} f(y;\theta)\,dy\right] = \beta. \tag{9.27}$$

The endpoints of a prediction interval $J(\mathbf{X})$ given by (9.27) are also called the *β-expectation tolerance limits* (Guttman, 1970).

When $X_1, X_2, \ldots, X_n$ and Y are iid IG(μ, λ), the exponent, except for the factor $(-\frac{1}{2})$, of their joint distribution can be expressed as the sum of three independent chi-square variables as follows:

$$\frac{\lambda}{\mu^2}\left[\sum\frac{(X_i - \mu)^2}{X_i} + \frac{(Y - \mu)^2}{Y}\right] = \lambda\sum\left(\frac{1}{X_i} - \frac{1}{\bar{X}}\right)$$

$$+ \frac{n\lambda(\bar{X} - \mu)^2}{\mu^2\bar{X}} + \frac{\lambda(Y - \mu)^2}{\mu^2 Y}. \tag{9.28}$$

The three terms on the right side in (9.28) are independently distributed as χ^2_{n-1}, χ^2_1 and χ^2_1, respectively. Next, the last two terms can be combined differently so that the right side equals

$$\lambda \sum \left(\frac{1}{X_i} - \frac{1}{\bar{X}} \right) + \frac{n\lambda(\bar{X} - Y)^2}{\bar{X}Y(n\bar{X} + Y)} + \frac{\lambda[(n\bar{X} + Y) - (n + 1)\mu]^2}{\mu^2(n\bar{X} + Y)},$$

(9.29)

where again the last two terms are independently distributed, each as χ^2_1. This independence can be easily established by finding the conditional moment-generating function of $n\lambda(\bar{X} - Y)^2 / \bar{X}Y(n\bar{X} + Y)$ given $n\bar{X} + Y = n\bar{x} + y$. This result also follows from the "analysis of reciprocals" property discussed in Chapter 6.

In (9.29) the first two terms do not involve the parameter μ and can be used to obtain 100β percent prediction intervals for Y when μ is unknown. When λ is assumed known, we can use the statistic

$$U = \frac{n\lambda(\bar{X} - Y)^2}{\bar{X}Y(n\bar{X} + Y)}.$$

(9.30)

Since $U \sim \chi^2_1$, it follows from $P[U \leqslant \chi^2_{1,\beta}] = \beta$, where $\chi^2_{1,\beta}$ is the 100β percentage point of the chi-square distribution with 1 df, that the 100β percent prediction interval for Y is given by

$$\left[\left(\frac{1}{\bar{X}} + \frac{1}{2\lambda} \chi^2_{1,\beta} \right) \pm \left\{ \frac{n+1}{n\lambda\bar{X}} \chi^2_{1,\beta} + \frac{1}{4\lambda^2} (\chi^2_{1,\beta})^2 \right\}^{1/2} \right]^{-1}.$$

(9.31)

When λ is unknown, consider the statistic W obtained by taking $(n - 1)$ times the ratio of first two terms of (9.29), so that

$$W = \frac{(n - 1)(\bar{X} - Y)^2}{\bar{X}Y(n\bar{X} + Y)V},$$

(9.32)

with $V = \sum(1/X_i - 1/\bar{X})/n$. W does not depend on λ (or μ) and has the $F_{1,n-1}$ distribution with 1 and $(n - 1)$ df. Hence, it follows from $P[W \leqslant F_{1,n-1,\beta}] = \beta$, where $F_{1,n-1,\beta}$ is the 100β percentage point of the $F_{1,n-1}$ distribution, that the 100β percent prediction interval for Y is given by

$$
\left\{ \left(\frac{1}{\bar{X}} + \frac{nV}{2(n-1)} F_{1,n-1,\beta} \right) \right.
$$

$$
\left. \pm \left[\frac{(n+1)V}{(n-1)\bar{X}} F_{1,n-1,\beta} + \frac{n^2 V^2}{4(n-1)^2} F^2_{1,n-1,\beta} \right]^{1/2} \right\}^{-1}. \qquad (9.33)
$$

In case of a nonpositive difference of the two terms in (9.31) or (9.33), the upper limit is easily seen to be ∞.

When μ is assumed known, then the two independent variables on the left side of (9.28) can be used directly. Clearly, the statistic $R = n(Y - \mu)^2 / YQ \sim F_{1,n}$, where $Q = \sum (X_i - \mu)^2 / X_i$. Again considering $P[R \leqslant F_{1,n,\beta}] = \beta$, it follows that the 100β percent prediction interval is given by

$$
\left(\mu + \frac{Q}{2n} F_{1,n,\beta} \right) \mp \frac{Q}{2n} \left[\frac{4n\mu}{Q} F_{1,n,\beta} + F^2_{1,n,\beta} \right]^{1/2}. \qquad (9.34)
$$

The statistics U, W, and R do not provide one-sided limits, except in special cases where there is a nonpositive difference of the two terms in (9.31) or (9.33) since two values of Y correspond to one value of each of these statistics. One may think of considering $\sqrt{U}$, $\sqrt{W}$, and $\sqrt{R}$ so that there is a one-to-one correspondence between values of Y and those of a statistic. However, the distribution of each of these latter statistics depends on the unknown parameter(s). For example, it follows from Theorem 4 in Folks and Chhikara (1978) that the distribution of statistic $W' = \sqrt{W}$, where

$$
W' = \frac{\sqrt{n-1}(\bar{X} - Y)}{[\bar{X}Y(n\bar{X} + Y)V]^{1/2}}, \qquad (9.35)
$$

obtained by letting $n_1 = n$ and $n_2 = 1$ in equation (19) of their article, has the density function of form,

$$
h(w'; \mu, \lambda) = [1 - a(w', \lambda/\mu)]g_{t,n-1}(w'), \qquad (9.36)
$$

where $g_{t,n-1}(w')$ is the density function of Student's t with $(n-1)$ df and $a(w', \lambda/\mu)$ depends on the ratio λ/μ of the unknown parameters. Similarly the distributions of the other two statistics,

$$\sqrt{U} = \frac{\sqrt{n\lambda}(\bar{X} - Y)}{[\bar{X}Y(n\bar{X} + Y)]^{1/2}}$$

and

$$\sqrt{R} = \frac{\sqrt{n}(Y - \mu)}{\sqrt{YQ}},$$

depend on the unknown parameters μ and λ, respectively. Thus the preceding approach is limited to providing exact prediction limits only for the two-sided case.

Padgett (1982) proposed a method to obtain approximate prediction intervals for the mean of future observations from IG. Suppose $Y_1, Y_2, \ldots, Y_m$ are the m random observations from the same distribution from which the first sample was drawn. Then

$$\frac{m\lambda(\bar{Y} - \mu)^2}{\mu^2\bar{Y}} \sim \chi_1^2,$$

where $\bar{Y} = \sum_1^m Y_i/m$. Since $n\lambda V \sim \chi_{n-1}^2$ and is independent of the future observations, the ratio statistic,

$$\frac{m(n-1)(\bar{Y} - \mu)^2}{n\mu^2\bar{Y}V} \sim F_{1,n-1}.$$

A 100β percent prediction interval for $\bar{Y}$ can be obtained by solving the inequality,

$$\frac{(\bar{Y} - \mu)^2}{\bar{Y}V} \leqslant \frac{n\mu^2 F_{1,n-1,\beta}}{m(n-1)}. \tag{9.37}$$

However, if μ is unknown, (9.37) cannot be used for obtaining an exact prediction interval. One may consider replacing μ by its estimate $\bar{X}$ and thereby obtain an approximate prediction interval. Padgett (1982) tried it and found by using Monte Carlo studies that such an approximation provided poor coverage probabilities. However, he found that if μ in the left side of (9.37) is replaced by the pooled sample mean, $\hat{\mu} = (n\bar{X} + m\bar{Y})/(m + n)$ so that

$$\frac{(\bar{Y} - \mu)^2}{\bar{Y}} \approx \left(\frac{n}{m + n}\right)^2 \left(\bar{Y} + \frac{\bar{X}^2}{\bar{Y}} - 2\bar{X}\right)$$

and μ^2 in the right side of (9.37) is replaced by $\bar{X}^2$, the resulting solution provides an approximate 100β percent prediction interval which has a much better coverage probability, that is, closer to β. Making the preceding substitutions in (9.37) and simplifying it further, the 100β percent prediction limits $l(x)$ and $u(x)$ are given by the roots of the quadratic equation

$$\bar{Y}^2 - C(\mathbf{X})\bar{Y} + \bar{X}^2 = 0, \tag{9.38}$$

where

$$C(\mathbf{X}) = \left[\frac{(n + m)^2}{nm(n - 1)}\right]\bar{X}^2 VF_{1,n-1,\beta} + 2\bar{X}.$$

Example For a comparison of the exact prediction limits given by (9.33) and the approximate ones by (9.38), we give these limits for a future repair time using the data set given earlier in Section 9.2.

For the observed repair times, $n = 46$, $\bar{x} = 3.61$, and $v = .587$; these are the respective MLEs of μ and $1/\lambda$. Suppose $\beta = .95$ and

Table 9.1 The Two-Sided 95 Percent Prediction Limits for a Future Repair Time

Type	Lower limit	Upper limit
Exact	0.33	49.10
Approximate	0.32	40.84
Bayesian	0.26	20.40

Table 9.2 Average Widths and Coverage Probabilities for Several Two-Sided Prediction Limits When $\mu = 1$

λ	β	n	Average width (Coverage probability)			
			Exact[a]	Bayesian[a]	MLE	Approximate
.25	.95	30	Infinite	8.6	7.4 (.92)	21.8 (.96)
	.99	30	Infinite	27.6	16.7 (.97)	38.4 (.99)
4	.90	15	2.1	1.7	1.6 (.86)	2.03 (.89)
	.95	30	2.4	2.0	1.9 (.93)	2.4 (.94)
	.99	15	4.6	3.5	2.8 (.97)	4.1 (.99)
	.99	30	3.7	3.0	2.7 (.98)	3.5 (.99)

[a]The coverage probability for the Bayesian limits is equal to β, but it generally exceeds β in the case of exact limits.

$m = 1$. Then the exact prediction limits obtained from (9.33) and the approximate prediction limits obtained by solving (9.38) are given in Table 9.1. Also, given in the table are the Bayesian prediction limits discussed earlier in Section 7.5 and which were obtained by solving numerically the two integrals in (7.37) with $(1 - \alpha)$ replaced by $\beta = .95$.

It is seen from Table 9.1 that the Bayesian limits have the shortest interval and the exact limits have the widest interval. The approximate limits appear to do well.

To make a fair comparison with the exact limits, the coverage probability of the approximate prediction limits should also be evaluated. In Table 9.2 we give certain results on the average width of the prediction interval and the associated coverage probability, when $\mu = 1$, $\lambda = .25$, 4, $n = 15$ and/or 30, and $\beta = .95$, .99. Given also are the average width of the MLE prediction intervals, which are easily obtained from solving

$$\int_0^{l(\mathbf{x})} f(y|\mathbf{x})dy = \frac{1 - \beta}{2} \quad \text{and} \quad \int_{u(\mathbf{x})}^{\infty} f(y|\mathbf{x})dy = \frac{1 - \beta}{2} \qquad (9.39)$$

for the lower and upper limits $l(\mathbf{x})$ and $u(\mathbf{x})$; here $f(y|\mathbf{x})$ is the MLE of the pdf $f(y)$, sometimes known as the estimative density function (Aitchison and Dunsmore, 1975).

Table 9.2 shows that the MLE limits have the smallest average width, but they also have the lowest probability coverage. The approximate limits compare fairly well in probability coverage (their coverage is almost equal to the nominal level), but these limits are superior to exact limits because of their shorter average width. The Bayesian limits are the best when considered in terms of both the average width and the probability coverage. This, of course, is expected since these prediction limits are obtained as the central 100β percent limits.

9.5 FAILURE RATE

The failure rate of a mechanism at time t is defined by the conditional probability that it fails during the infinitesimal time interval $(t, t + h)$ given that no failure occurred before t. If $f(t)$ is the density function of the mechanism failure time, its failure rate $r(t)$ at time t is therefore given by

$$r(t) = \frac{f(t)}{1 - F(t)}, \qquad t > 0, \tag{9.40}$$

where $F(t)$ is the cumulative distribution function. For the inverse Gaussian distribution, $IG(\mu, \lambda)$, the denominator $1 - F(t) = R(t)$ and is given in (9.1). Thus the failure rate of IG is

$$r(t) = \frac{(\lambda 2\pi t^3)^{1/2} \exp(-\lambda(t - \mu)^2/2\mu^2 t)}{\Phi(\sqrt{\lambda/t}(1 - t/\mu)) - e^{2\lambda/\mu}\Phi(-\sqrt{\lambda/t}(1 + t/\mu))},$$

$$t > 0. \tag{9.41}$$

The expression for $r(t)$ is rather complicated but it is not difficult to compute for given μ and λ. Several typical failure rate curves are given in Figure 9.3. Inspection of these curves makes it obvious that the failure rate is not monotonic for all μ and λ. However, one might be led to ask whether it is monotonic for some parameter values. We shall show that $r(t)$ is nonmonotonic.

The mode of the inverse Gaussian occurs at

$$t_m = -\frac{3\mu^2}{2\lambda} + \mu\left(1 + \frac{9\mu^2}{4\lambda^2}\right)^{1/2}. \tag{9.42}$$

Because $f(t)$ is increasing and $R(t)$ is decreasing as t increases from 0 to t_m, it follows that $r(t)$ is increasing for $t \leqslant t_m$ and it achieves the maximum value at some point $t^* \in (t_m, \infty)$. Differentiating $r(t)$, it follows that the derivative

$$r'(t) = r(t)\left[\frac{f'(t)}{f(t)} + \frac{f(t)}{1 - F(t)}\right]. \tag{9.43}$$

Let

$$p(t) = \frac{-f'(t)}{f(t)}. \tag{9.44}$$

Since $f(t)$ is decreasing for $t > t_m$, $p(t) > 0$ and we find that

$$p(t) = \frac{\lambda}{2\mu^2} + \frac{3}{2t} - \frac{\lambda}{2t^2}. \tag{9.45}$$

It is easily seen that $p(t)$ increases as t increases from 0 to $t_0 = 2\lambda/3$, and then decreases, approaching $\lambda/2\mu^2$ asymptotically as $t \to \infty$.

We now show that $r(t)$ achieves its maximum value at some point $t^* \in [t_m, t_0]$. Rewriting (9.43), we have

$$\frac{r'(t)}{r(t)} = \frac{p(t)}{1 - F(t)} \left[\int_t^\infty \frac{f'(x)}{p(x)} dx + \frac{f(t)}{p(t)} \right]. \tag{9.46}$$

For $t > t_0$, $p(t)$ is decreasing and

$$\int_t^\infty \frac{f'(x)}{p(x)} dx + \frac{f(t)}{p(t)} < 0.$$

Thus it follows from (9.46) that $r'(t) < 0$ for $t > t_0$, implying $r(t)$ decreases as t increases from t_0 to ∞.

Note that $r(t)$ is increasing for $t \leqslant t_m$ and decreasing for $t > t_0$. Because it is continuous and differentiable, $r'(t)$ must vanish somewhere in the interval $[t_m, t_0]$. From (9.43), we have

$$r'(t) = r(t)[-p(t) + r(t)]$$

and $r'(t) = 0$ for $t \in [t_m, t_0]$ if and only if $r(t) = p(t)$. Since $p(t)$ is a monotonic increasing function for $t \in [t_m, t_0]$, and in the same interval $r'(t)$ vanishes at some point, say t^*, obviously $p(t)$ intersects $r(t)$. Since their point of intersection satisfies $r'(t) = 0$, $p(t)$ intersects $r(t)$ at t^*. Next, for the function $p(t)$, $p(\lambda/3) = \lambda/2\mu^2$, where $t_m < \lambda/3 < t_0$, and after approaching a maximum value at t_0, $p(t)$ approaches $\lambda/2\mu^2$, which is also the limit $r(t)$ reaches as $t \to \infty$ as shown below. Therefore, $p(t)$ intersects $r(t)$ at t from below and $r'(t) = 0$ has a single root, t^*. Hence, $r(t)$ is nonmonotonic; it first increases, attains a maximum value at t^*, where

$$\frac{\lambda}{3} < t^* < \frac{2\lambda}{3}, \tag{9.47}$$

and then decreases. The modal value t^* is the solution of the equation

$$r(t) = \frac{\lambda}{2\mu^2} + \frac{3}{2t} - \frac{\lambda}{2t^2} \tag{9.48}$$

and the maximum value of $r(t)$ is $r(t^*)$.

Examination of the graphs in Figure 9.3 indicates there exists a nonzero asymptotic value of $r(t)$ unlike the failure rate of the lognormal, which approaches zero asymptotically as $t \to \infty$. Making use of the L'Hospital's rule, it is seen that

$$\lim_{t \to \infty} r(t) = \lim_{t \to \infty} \frac{f'(t)}{-f(t)} = \lim_{t \to \infty} p(t). \tag{9.49}$$

Because of (9.45), the last limit in (9.49) is $\lambda/2\mu^2$. Hence $r(t)$ is asymptotically equal to $\lambda/2\mu^2$ at $t \to \infty$.

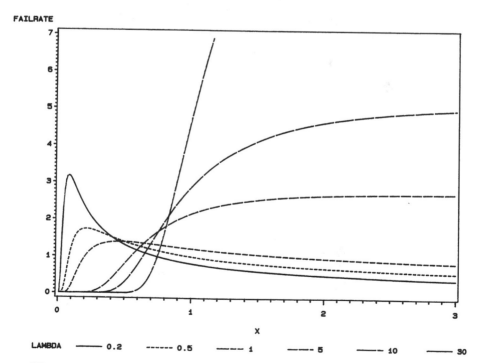

Figure 9.3 IG failure rate with $\mu = 1$ for six values of λ.

From Figure 9.3 one can observe that $r(t)$ is virtually nondecreasing for all t when λ is large relative to μ. This can also be seen analytically. When λ is large, t^*, the point where $r(t)$ is maximum, falls into the extreme right tail of the distribution. Moreover $r(t)$ becomes infinite as $\lambda \to \infty$ and $t \to \infty$ and the distribution has an increasing failure rate (IFR).

9.6 MEAN RESIDUAL LIFE

Sometimes manufactured products are initially tested for a short time period before being marketed. This initial testing is to detect defective items and to safeguard against business losses that may result from marketing a bad product. Thus the useful life of an initially tested product is the length of time given by its total survival time less the time period of its initial test. There are other situations as well where one may be interested in studying the remaining life or, what we call, the residual life of a lifetime phenomenon.

Let S be the residual lifetime consisting of the period from time T until the time of failure given that there was no failure prior to T. Then the density function of S, $g_T(t)$, is given by

$$g_T(t) = \frac{f(t)}{R(T)}, \qquad t > T \tag{9.50}$$

and the mean residual lifetime,

$$\mu_T = \frac{1}{R(T)} \int_T^\infty (t - T) f(t)\, dt. \tag{9.51}$$

Making the transformation $y = \sqrt{\lambda}(t - \mu)/\mu\sqrt{t}$ as in Theorem 2.1 and simplifying the first integral in (9.51), it easily follows that

$$\int_T^\infty t\, f(t)\, dt = \mu \left[\Phi\left(\sqrt{\frac{\lambda}{T}}\left(1 - \frac{T}{\mu}\right)\right) \right.$$
$$\left. + e^{2\lambda/\mu} \Phi\left(-\sqrt{\frac{\lambda}{T}}\left(1 + \frac{T}{\mu}\right)\right) \right].$$

Since the second term on the right side in (9.51) is equal to $-T$, the expression in (9.51) after it is simplified by substituting for $R(T)$ from (9.1), is given by

$$\mu_T = \frac{(\mu - T)\Phi\left(\sqrt{\frac{\lambda}{T}}\left(1 - \frac{T}{\mu}\right)\right) + (\mu + T)e^{2\lambda/\mu}\Phi\left(-\sqrt{\frac{\lambda}{T}}\left(1 + \frac{T}{\mu}\right)\right)}{\Phi\left(\sqrt{\frac{\lambda}{T}}\left(1 - \frac{T}{\mu}\right)\right) - e^{2\lambda/\mu}\Phi\left(-\sqrt{\frac{\lambda}{T}}\left(1 + \frac{T}{\mu}\right)\right)}.$$

(9.52)

Considering a different formulation [e.g., see Watson and Wells (1961)], the mean residual lifetime can be expressed in terms of the failure rate as

$$\mu_T = \int_0^\infty \exp\left(-\int_T^{t+T} r(x)\,dx\right)dt.$$

(9.53)

Clearly, μ_T decreases (increases) monotonically for the lifetime distribution with IFR (DFR). However, if $r(t)$ first increases, and then begins to decrease monotonically at some time t^*, as is the case with the inverse Gaussian, we have $\mu_{T_2} > \mu_{T_1}$ for $T_2 > T_1 > t^*$. Thus the mean residual lifetime for the inverse Gaussian increases from some time onward. The increase in the mean residual lifetime can be justified in terms of the physical behavior of a product. For example, if the items with a high rate of defect or wear are excluded after these have failed, the rate of failure of the surviving items will be relatively small and their average lifetime will result in a higher longevity than all items combined. Since the inverse Gaussian distribution has the nonmonotonic failure rate property that can characterize the dominance of early occurrences, it is highly suitable for a failure model to study the mean residual life of a manufactured product with a higher rate of defective items in the beginning.

To examine the asymptotic behavior of μ_T as $T \to \infty$, consider repeated use of L'Hospital's rule for (9.52), since it becomes indeterminate otherwise. Consequently, it follows that

$$\lim_{T\to\infty} \mu_T = \lim_{T\to\infty}\left[\frac{3}{2T} + \frac{\lambda(T-\mu)}{\mu^2 T} - \frac{\lambda(T-\mu)^2}{2\mu^2 T^2}\right]^{-1} = \frac{2\mu^2}{\lambda}.$$

On the other hand, it is easy to see from (9.52) that $\mu_T \to \mu$ as $T \to 0$. Accordingly, we have the following result:

For a fixed μ and λ, (i) $\mu_T/\mu \to 1$ as $T \to 0$ and (ii) $\mu_T/\mu \to 2\mu/\lambda$ as $T \to \infty$.

Thus the mean residual lifetime will eventually exceed the mean lifetime whenever $\mu > \lambda/2$.

9.7   INVERSE GAUSSIAN VERSUS OTHER DISTRIBUTIONS AS LIFETIME MODELS

In reliability studies the choice of distribution is often made on the basis of what is understood about the failure mechanism. For example, it is logical to consider an increasing failure rate (IFR) distribution for representing lifetime in situations which mainly involve aging or the wearing-out process. However, failure may be due to various other causes such as technological defect resulting from violation of the system design, improper usage, instantaneous injury, and so on (Gertsbakh and Kordonskiy, 1969). Thus it is more appropriate to consider the physical characteristics of a failure phenomenon than the goodness of data fit by a distribution to make a choice of failure model. Lawless (1983) also emphasizes this point in selection of a model and suggests that model-dependent analyses need to be scrutinized for robustness and so on by a careful examination of assumptions and data. For example, he shows that though both the lognormal and Weibull models also provide good fit for the data given in Section 5.6, there is a huge disparity in the 95 percent confidence intervals for the .01 quartile under the two models. Easterling (1976) examines the role of model fitting, but he, of course, does it from the viewpoint of parametric estimation. His ideas may be relevant here.

When early occurrences such as product failures or repairs are dominant in a lifetime distribution, its failure rate is expected to be nonmonotonic, first increasing and later decreasing. In that situation the inverse Gaussian distribution might provide a suitable choice for a lifetime model. Though the lognormal distribution, among others, is also applicable in such cases, there are certain advantages in choosing the inverse Gaussian over the lognormal. First of all, the inverse Gaussian addresses a wider class of lifetime distributions. For example, as shown in Section 9.5, the inverse Gaussian is almost an IFR distribution when it is slightly skewed, and hence is also applicable to describe lifetime distribution which is not dominated by early failures. Secondly, the failure rate $r(t)$ is nonzero and constant as $t \to \infty$ for the inverse Gaussian, but $r(t)$ goes to 0 as $t \to \infty$ for the lognormal. The nearly constant failure rate after a certain time period implies that the occurrence of failure is purely random and is independent of past life; this is a property of the failure rate of an exponential distribution which has been extensively used in reliability studies. On the other hand, a near zero failure rate implies that almost no failure will occur, which is hardly feasible in real life. For the repair times of an airborne communication transceiver given in Section 9.2, Chhikara and Folks (1977) discussed further the differences between the inverse Gaussian and the lognormal and made a case for the IG over the lognormal for a lifetime model.

Folks and Chhikara (1978) fitted the inverse Gaussian distribution to a variety of data sets. Among others, data on shelf life of food product M given in Section 6.3 was considered and the inverse Gaussian fit was shown to be as good as the lognormal and Weibull distribution fits previously considered by Gacula and Kubala (1975). In his discussion of the paper by Folks and Chhikara (1978), Professor Aitkin fitted two other three-parameter families of distributions, the generalized gamma and the transformed normal, to the shelf life data and showed that the inverse Gaussian distribution competed well in this application with these three-parameter distributions.

10

Applications

10.1 INTRODUCTION

In this chapter we review practical applications of the inverse Gaussian—some of which are data oriented. Based on its origin of Brownian motion it is natural that one would think first of applications from the natural and physical sciences. The first example is from cardiology; the second from hydrology.

We also review some other applications outside the natural sciences. When considering human behavior, it may seem farfetched at first to postulate the existence of an underlying Brownian motion and, consequently, a first passage time. However, as we shall show by several examples, this is not at all unreasonable. Moreover, the use of the inverse Gaussian distribution is justified by goodness-of-fit considerations.

Although it is appealing to base the use of the IG distribution upon an underlying Wiener process, it is not at all critical. In the early days of the normal distribution, it was considered primarily as a law of errors, and the variable was considered to be the sum of many independent errors. Gradually, it became acceptable to use the normal distribution to describe all sorts of data. The situation with the IG distribution seems to be similar.

10.2   TRACER DYNAMICS

When a substance is injected into the bloodstream, the concentration of the substance can be monitored for some time afterward. Such data allows estimation of the probability distribution for the time that a particle of the substance remains in the blood. Sheppard and Savage (1951) and Sheppard (1962) made use of the inverse Gaussian distribution, among others, to describe the time distribution. Wise (1971, 1975) and Wise et al. (1968) also developed the inverse Gaussian as a possible model to describe cycle time distribution for particles in the blood.

Following injection of a radionuclide such as ^{47}Ca, the specific activity can be estimated over a period of time and a curve fitted to the data. Wise et al. (1968) fitted a gamma curve to the data. When normalized, it can be interpreted as a probability distribution of the time that a particle remains in the blood. From this, they deduced that the inverse Gaussian distribution describes the cycle times from plasma to nonplasma. The model was developed in the following way: Once a particle is injected into the blood, it goes through independent cycles of passing to the tissue and back to the blood until it is excreted or deposited in bone for a long period. Let p be the probability of completing a cycle and $q = 1 - p$. Then the number of cycles completed is a geometric variable N, with probability function

$$P_N(n) = p^n q, \qquad n = 0, 1, 2, \ldots, 0 < p < 1, q = 1 - p. \qquad (10.1)$$

Consider the time T from injection of a particle into the bloodstream until it is excreted or deposited in bone. Denote its density function by $f_T(t)$ and denote the conditional distribution of T, given n cycles are completed, by

$$f_{T|N}(t \mid n), n = 0, 1, 2, \ldots, \qquad t > 0. \qquad (10.2)$$

Then the unconditional distribution of T is given by

$$f_T(t) = \sum_{n=0}^{\infty} p^n q f_{T|N}(t \mid n).$$ (10.3)

The density function $f_{T|N}(t \mid n)$ for $n \geqslant 1$ is considered to be the distribution of the sum of n independent and identically distributed variables, each with density function $g(t)$, known as the cycle time distribution. Note that when there are n cycles, the total time T is the sum of $n + 1$ variables: n independent cycle times $T_1, T_2, \ldots, T_n$ and T_0, the time from entry into the blood to excretion. Ignoring T_0, the density $f_T(t)$ is given approximately by excluding the term $n = 0$. Wise et al. justify this approximation by noting that a few minutes after injection, the distribution of cycle times takes over in importance. Actually, of course, (10.3) does not give a density when the summation excludes $n = 0$ but would need to be divided by a norming constant. In the derivation that follows, that same norming constant would appear in all of the moments and cumulants. However, because the effect is minimal and the argument is approximate in the end, we shall follow Wise in ignoring the constant.

Wise assumed a gamma distribution for $f_T(t)$. Using the cumulants of the gamma and equation (10.3), he obtained the cumulants of $f_{T|N}$ and in the limit as $p \to 1$, the cumulant-generating function of $g(t)$, the cycle time distribution. Denote the moments and cumulants of f_T by μ_r and L_r, respectively. Denote the moments and cumulants of $g(t)$ by v_r and K_r, respectively. Using equation (10.3) and excluding $n = 0$,

$$\mu_r = \int_0^{\infty} t^r f_T(t)\, dt$$

$$= q \sum_{n=1}^{\infty} p^n \int_0^{\infty} t^r f_{T|N}(t \mid n)\, dt.$$ (10.4)

We can express the moments of f_T in terms of its cumulants L_r, which then can be expressed in terms of the cumulants of g, K_r.

$$\mu_1 = qK_1 \sum_{n=1}^{\infty} np^n,$$

$$\mu_2 = q \sum_{n=1}^{\infty} p^n(nK_2 + n^2K_1^2),$$

$$\mu_3 = q \sum_{n=1}^{\infty} p^n(nK_3 + 3n^2K_2K_1 + n^3K_1^3),$$

$$\mu_4 = q \sum_{n=1}^{\infty} p^n(nK_4 + 4n^2K_3K_1 + 6n^3K_2K_1^2 + n^4K_1^4),$$

etc.   (10.5)

Wise then chose to work in terms of polynomials in p appearing in equation (10.5). Let

$$S_0 = p(1 - p)^{-1},$$
$$S_1 = p(1 - p)^{-2},$$
$$S_2 = (p + p^2)(1 - p)^{-3},$$
$$S_3 = (p + 4p^2 + p^3)(1 - p)^{-4},$$
$$S_4 = (p + 11p^2 + 11p^3 + p^4)(1 - p)^{-5},$$

etc.   (10.6)

The moments μ_r and, consequently, the cumulants L_r can be expressed in terms of these quantities. Then solving for the cumulants in terms of q and p, we obtain:

$$L_1 = K_1q^{-1},$$
$$L_2 = K_1^2pq^{-2} + K_2q^{-1},$$
$$L_3 = K_1^3(p + p^2)q^{-3} + 3K_1K_2pq^{-2} + K_3q^{-1},$$
$$L_4 = K_1^4(p + 4p^2 + p^3)q^{-4} + 6K_2K_1^2(p + p^2)q^{-3}$$
$$+ (4K_3K_1 + 3K_2^2)pq^{-2} + K_4q^{-1},$$

$$L_5 = K_1^5(p + 11p^2 + 11p^3 + p^4)q^{-5}$$
$$+ 10K_2K_1^3(p + 4p^2 + p^3)q^{-4}$$
$$+ (10K_3K_1^2 + 15K_2^2K_1)(p + p^2)q^{-3}$$
$$+ (5K_4K_1 + 10K_3K_2)pq^{-2} + K_5q^{-1},$$

etc. (10.7)

Finally let us assume that the f_T is a gamma distribution with the following density function

$$f_T(t) = \frac{\beta^{1-\alpha}t^{-\alpha}e^{-\beta t}}{(-\alpha)!}.$$ (10.8)

Then the cumulants L_r are given by

$$L_r = (1 - \alpha)(r - 1)\beta^{-r}.$$ (10.9)

If we substitute the values in (10.9) and (10.7) and solve for the K_r, we obtain, putting $\alpha' = 1 - \alpha$:

$$K_1 = \alpha'q\beta^{-1},$$
$$K_2 = \alpha'q(1 - \alpha' + \alpha'q)\beta^{-2},$$
$$K_3 = \alpha'q\beta^{-3}[(1 - \alpha')(2 - \alpha' + 3q\alpha') + 2q\alpha'^2],$$
$$K_4 = \alpha'q\beta^{-4}[(1 - \alpha')(2 - \alpha')(3 - \alpha')$$
$$+ q(1 - \alpha')(11\alpha' - 7\alpha'^2)$$
$$+ 12q^2(1 - \alpha')\alpha'^2 + 6q^3\alpha'^3)],$$
$$K_5 = \alpha'q\beta^{-5}[(1 - \alpha')(2 - \alpha')(3 - \alpha') + (4 - \alpha')$$
$$+ 5q\alpha'(1 - \alpha')(2 - \alpha')(5 - 3\alpha')$$
$$+ 10q^2\alpha'(1 - \alpha')(7 - 5\alpha')$$
$$+ 60q^3\alpha'^3(1 - \alpha') + 24q^4\alpha'^4],$$ (10.10)

etc.

For q very small, the cumulants K_r are given by

$$K_1 = \frac{\alpha' q}{\beta},$$

$$K_2 = \frac{\alpha' q(1 - \alpha')}{\beta^2},$$

$$K_3 = \frac{\alpha' q(1 - \alpha')(2 - \alpha')}{\beta^3}, \qquad (10.11)$$

etc.

So for q very small, the cumulant-generating function of $g(t)$ is given by

$$K(\theta) = q \left[\frac{\alpha' \theta}{\beta} + \frac{\alpha'(1 - \alpha')}{\beta^2} \frac{\theta^2}{2!} + \frac{\alpha'(1 - \alpha')(2 - \alpha')}{\beta^3} \frac{\theta^3}{3!} + \cdots \right]$$

$$= q \left[1 - \left(1 - \frac{\theta}{\beta} \right)^{1 - \alpha} \right]. \qquad (10.12)$$

Now it is apparent from (2.7) that for $\alpha = \frac{1}{2}$, the cumulant-generating function is that of an inverse Gaussian.

Wise and his colleagues do not restrict themselves to the case where $\alpha = \frac{1}{2}$, but consider the more general passage time distribution without this restriction.

It seems that one should be able to obtain the cumulant-generating function $K(\theta)$ more directly. The moment-generating function of f_T, say $M(\theta)$, would be given by

$$M(\theta) = E[e^{T\theta}]$$

$$= \int_0^\infty e^{t\theta} f_T(t) \, dt$$

$$= q \sum_{n=1}^\infty p^n \int_0^\infty e^{t\theta} f_{T|N}(t \mid n) \, dt. \qquad (10.13)$$

Remembering that T is the sum of n independent variables each with density function $g(t)$, with moment-generating function, say $N(\theta)$, we have the cumulant-generating function of f_T given by

$$L(\theta) = \log \left\{ q \sum_{n=1}^{\infty} p^n [N(\theta)]^n \right\}. \tag{10.14}$$

Equation (10.14) gives us a formal relationship between the cumulants of f_T and the moments of g. Differentiating $L(\theta)$, setting $\theta = 0$, and using the relationships between cumulants and moments, we can obtain equation (10.7) in a more straightforward fashion than was done by Wise. For example,

$$\frac{dL(\theta)}{d\theta} = \frac{\sum_{n=1}^{\infty} nqp^n N^{n-1}(\theta) \, dN(\theta)/d\theta}{\sum_{n=1}^{\infty} qp^n N^n(\theta)} \tag{10.15}$$

and therefore

$$L_1 = \frac{dL(\theta)}{d\theta}\bigg|_{\theta=0} = \frac{K_1(q/p)}{p} = \frac{K_1}{q}.$$

10.3 EMPTINESS OF A DAM

At about the time that Tweedie's extensive papers appeared on the inverse Gaussian distribution, there began a series of papers considering an infinite dam with a random input and a steady release. One of the problems considered was determination of the probability distribution of the time until the release stops for the first time. These first passage problems are similar to those given elsewhere in this work and are formulated as follows:

Suppose that water is flowing into a reservoir with infinite storage capacity. The continuous input $X(t)$ which enters during time t has an infinitely divisible distribution. The content at time t is denoted by a continuous variable $Z(t)$. The content at time zero, is $Z(0) = z_0$. Consider the random time at which the reservoir first runs dry, $T(z_0)$.

Following Gani and Prabhu (1963) or Kendall (1957), we can obtain the density function $g(z_0; t)$ of $T(z_0)$ by finding the Laplace transform

$$g*(\theta; z_0) = \int_0^\infty e^{-\theta t} g(z_0; t)\, dt. \tag{10.16}$$

Now the Laplace transform has the form

$$g*(\theta; \ z_0) = \exp[-z_0\eta(\theta)], \tag{10.17}$$

where $\eta(s)$ satisfies the functional equation

$$\eta(\theta) = \theta + \xi[\eta(\theta)]. \tag{10.18}$$

where $\eta(\theta) > 0$ for θ real and positive.

Hasofer (1964) considered the situation where the input $X(t)$ is inverse Gaussian with $\mu = \alpha t$ and $\lambda = \beta t^2$, $\alpha > 0$, $\beta > 0$. Actually, Hasofer used a rather different notation but we are adhering to Tweedie's parameterization, as we have done throughout. The Laplace transform defined by $E[e^{-\theta X}]$ is then found explicitly and

$$\xi(\theta) = \frac{\beta}{\alpha}\left(\sqrt{1 + \frac{2\alpha^2\theta}{\beta}} - 1\right). \tag{10.19}$$

It follows from Hasofer's development that when $\alpha < 1$, $P(T(z_0) < \infty) = 1$ so we shall assume for the time being that $\alpha < 1$. Using equation (10.19) in equation (10.18) leads to a defining quadratic equation for $\eta(\theta)$. For θ real and positive, only one of the solutions satisfies the condition that $\eta(\theta) > 0$, so

$$\eta(\theta) = \theta + \beta\frac{(\alpha - 1)}{\alpha} + \sqrt{\beta^2\frac{(\alpha - 1)^2}{\alpha} + 2\beta\theta}. \tag{10.20}$$

Then from (10.17) and (10.20) and recalling that $\alpha < 1$ we have the moment-generating function

$$m_{T(z_0)}(\theta) = E[e^{\theta T(z_0)}]$$

$$= \exp\left\{\theta z_0 + \beta z_0\left(\frac{1-\alpha}{\alpha}\right)\left[1 - \sqrt{1 - 2\frac{\theta}{\beta}\left(\frac{\alpha}{\alpha-1}\right)^2}\right]\right\}. \tag{10.21}$$

This we recognize as the moment-generating function of a translated inverse Gaussian variable. That is

$$T(z_0) - z_0 \sim \text{IG}\left(\frac{\alpha z_0}{1 - \alpha}, \beta z_0^2\right). \tag{10.22}$$

Next Hasofer obtains the same distribution for $T(z_0)$, regardless of whether α (ρ in his notation) is > 1 or < 1. However, if we allow α to exceed 1, we no longer obtain a density function. Rather we obtain the conditional density of $T(z_0)$ given that $T(z_0) < \infty$ multiplied by the probability that $T(z_0) < \infty$. That is, if we let α exceed 1, we obtain what Whitmore (1979) calls the defective inverse Gaussian and it follows that

$$P(T(z_0) < \infty) = e^{2\lambda/\mu}$$
$$= e^{2\beta z_0(1 - \alpha)/\alpha}. \tag{10.23}$$

Although these results are quite intriguing, there is no indication that they match the behavior of any real dam. The main stimulus for the entire series of papers seems to have been based upon theoretical appeal rather than practical utility.

10.4 A PURCHASE INCIDENCE MODEL

Banerjee and Bhattacharyya (1976) presented a very interesting use of the inverse Gaussian distribution and a natural conjugate distribution. Data were available on the family purchase of toothpaste in the Chicago area from January 1958 to April 1963. It was postulated that for each family the interpurchase times follow an inverse Gaussian distribution. This conjecture seems reasonable when we think of the tube capacity as the barrier. Then if toothpaste usage follows Brownian motion with positive drift, the time until the next purchase should be an inverse Gaussian variable. The difference among the households in the population are modeled by the natural conjugate distribution on the IG parameters.

The data used in this study were from 289 families. Since different size tubes were purchased, the data were scaled as follows: If a nine ounce tube lasted five weeks, it was recorded as having an interpurchase time of 5/9. In other words, the data consist of the usage time per ounce of toothpaste for individual households.

Individual IG distributions were fitted to data from 12 households making at least 100 purchases in the five-year period. In no case was the goodness-of-fit χ^2 statistic significant at the 5 percent level. In fact the observed significance levels ranged from 0.053 to 0.634. The distributions were parameterized in terms of $\psi = 1/\mu$ and λ and were fitted by using maximum likelihood estimates of ψ and λ. Because of the good fits obtained on the families for which considerable data were available, it was decided to fit an IG distribution to each of the 289 families.

To describe the heterogeneity in the population a bivariate distribution was fitted to the 289 $(\hat{\psi}, \hat{\lambda})$ pairs. The natural conjugate distribution for this particular parameterization is given by

$$
f(\psi, \lambda) = \left(\frac{\beta}{\alpha}\right)^{1/2} \left(\frac{\gamma\alpha}{2}\right)^{\gamma/2} \left[H_\nu(\xi)B\left(\frac{\nu}{2}, \frac{1}{2}\right)\Gamma\left(\frac{\gamma}{2}\right)\right]^{-1}
$$

$$
\times \exp\left\{-\frac{\gamma\alpha}{2}\left[1 + \frac{\beta}{\alpha}\left(\psi - \frac{1}{\beta}\right)^2\right]\lambda\right\}\lambda^{(\gamma/2)-1}, \quad (10.24)
$$

where $\nu = \gamma - 1$, $\xi = (\alpha\beta/\nu)^{-1}$, and $H_\nu(\cdot)$ is the cumulative distribution function of Student's t-distribution with ν degrees of freedom. This distribution was fitted by the method of moments. A visual inspection of the observed and fitted frequencies as well as a chi-square test indicated a good fit.

The marginal distributions of ψ and λ are given by

$$
f(\psi) = \left(\frac{\beta}{\alpha}\right)^{1/2} \left[H_\nu(\xi)B\left(\frac{\nu}{2}, \frac{1}{2}\right)\right]^{-1} \left\{1 + \frac{\beta}{\alpha}\left(\psi - \frac{1}{\beta}\right)\right\}^{2-(\nu+1)/2}
$$

$$(10.25)$$

and

$$
f(\lambda) = \left(\frac{\gamma\alpha}{2}\right)^{\nu/2} \left[H_\nu(\xi)\Gamma\left(\frac{\nu}{2}\right)\right]^{-1} \Phi(z) \exp\left(\frac{-\gamma\alpha\lambda}{2}\right)\lambda^{(\nu/2)-1}
$$

where $z = (\gamma\lambda/\beta)^{1/2}$.

Again, a visual inspection of the observed and fitted frequencies, an inspection of the sample histograms and fitted curves, and chi-square goodness-of-fit tests indicated good fits.

The authors obtained the distribution of $N(t)$, the number of ounces of toothpaste purchases per household in time t. Using the parameter estimates for ψ, λ, and α, β, and γ, fitted values for $N(t)$ were obtained and compared with observed values. The agreement between these sets of values serves to justify their use of the IG and natural conjugate as a model.

10.5 THE DISTRIBUTION OF STRIKE DURATION

Lancaster (1972) made effective use of the inverse Gaussian distribution in describing strike duration data. Apart from the fact that the IG distribution gave a good fit to the data, the rationale for using the distribution was based on the idea of an underlying Wiener process. Lancaster postulated the existence of an underlying scalar measure of agreement between the two negotiating parties (actually he used a measure of difference). Then the duration of the strike depends on the measure of agreement $X(t)$. Starting from an agreement of x_0, the two parties go through a back and forth process gradually increasing the agreement until $X(t)$ finally reaches the point at which the strike is determined. It is not unreasonable to think of the process as being like Brownian motion with positive drift so that the strike duration is the first passage time.

Lancaster obtained data on duration of strikes in the United Kingdom between 1965 and 1972. These data were for eight industries: (1) metal manufacture, (2) nonelectrical engineering, (3) distribution trades, (4) vehicles, (5) construction, (6) ship building, (7) transport, and (8) electrical machinery. The data arose in grouped form. Further, because the Ministry of Labour statistics did not give the number of strikes lasting less than a day, the data are truncated. Because of these two difficulties, Lancaster used a numerical procedure to obtain maximum likelihood estimates for the truncated, grouped data. The maximum likelihood estimates of μ and λ for the eight industries are given in Table 10.1. The chi-square statistics for goodness of fit are also shown. As can be seen, the fit provided by the IG is good in all cases. Inspection of the

Table 10.1   Statistics from Eight Industries

Industry	n	$\hat{\mu}$	$\hat{\lambda}$	χ^2	df
1	198	7.229	2.670	12.4	13
2	149	5.076	1.924	5.8	10
3	54	4.329	2.058	3.1	2
4	103	3.922	1.520	1.3	9
5	225	7.463	3.968	13.1	11
6	112	6.061	2.732	8.3	9
7	102	1.012	0.119	4.0	4
8	72	4.292	0.837	16.2	4

parameter estimates suggests that industries 1 to 6 could be fitted by a single distribution. This was done subsequently and again the fit was quite good. Lancaster obtained $\hat{\mu} = 6.25$, $\hat{\lambda} = 2.627$, and $\chi^2 = 17.3$ with 22 degrees of freedom.

Of course, the analysis of residuals discussed in Section 6.4.1 gives an F test for testing that the IG distribution is the same for several populations. For the present example it is impossible to calculate exactly the necessary quantities because the data are grouped and truncated. As an approximation, however, we used $\hat{\mu}_i$ as the sample mean $\bar{x}_{i.}$ for the ith industry and we used $n_i/\hat{\lambda}_i$ as the sum of residuals $\sum_j (1/x_{ij} - 1/\bar{x}_{i.})$ for the ith industry. The calculations are shown in Table 10.2 for the first six industries as well as for all eight industries.

Thus in both cases we would conclude that the single distribution hypothesis should be rejected. This conclusion is at variance with that reached by Lancaster and should be scrutinized carefully for several reasons: (1) the way in which the quantities $\bar{x}$ and $\sum(1/x_i - 1/\bar{x})$ were obtained, (2) the unknown distribution of F for truncated data, and (3) the unknown distribution of F under the alternative distribution.

If the distributions have a common λ, the F statistic seems like a reasonable test for equality of means. However, if there is not a common λ, it is subject to the usual problems resulting from unequal variances in the normal case.

The Bartlett's test given in Section 6.4.1 can be applied to the

Table 10.2 Test of Homogeneity

Industry (i)	n_i	$\bar{x}_i$	$1/\bar{x}_i$	$\sum_j(1/x_{ij} - 1/\bar{x}_i)$
1	198	7.299	.137	74.157
2	149	5.076	.197	77.456
3	54	4.329	.231	26.234
4	103	3.922	.255	67.745
5	225	7.463	.134	56.701
6	112	6.061	.165	40.995
7	102	1.012	.988	854.865
8	72	4.292	.233	86.015

	First Six	All Eight
$\sum n_i(1/\bar{x}_{i.} - 1/\bar{x}_{..})$	7.761	77.743
$\sum_i\sum_j(1/x_{ij} - 1/\bar{x}_{i.})$	343.288	1284.168
$k - 1$	5	7
$\sum n_i - k$	835	1007
F	3.776	8.709
SL	$<.005$	$<.005$

data of Table 10.2 to test the hypothesis of a common λ. Recall that M/C is distributed as chi-square with $k - 1$ degrees of freedom where

$$M = f_. \log\left(\frac{v_.}{f_.}\right) - \sum f_i \log\left(\frac{v_i}{f_i}\right)$$

$$C = 1 + \frac{1}{3(k-1)}\left(\sum\frac{1}{f_i} - \frac{1}{\sum f_i}\right)$$

$$v_i = \sum_j\left(\frac{1}{x_{ij}} - \frac{1}{\bar{x}_{i.}}\right)$$

$$f_i = n_i - 1$$

$$v_. = \sum v_i$$

$$f_. = \sum f_i. \tag{10.26}$$

Table 10.3   Test for Equality of λ_i

Industry (i)	v_i	f_i	$f_i \log(v_i/f_i)$	$1/f_i$
1	74.157	197	−192.47279	.00508
2	77.456	148	−95.83033	.00676
3	26.234	53	−37.27149	.01887
4	67.745	102	−41.74067	.00980
5	56.701	224	−307.74334	.00446
6	40.995	111	−110.56489	.00901
Total	343.288	835	−785.62351	.05398

Using the figures in Table 10.2, the calculation of M/C proceeds as shown in Table 10.3

$$f.\log\left(\frac{v_.}{f_.}\right) = 835 \log\left(\frac{343.288}{835}\right) = -742.19975$$

$$M = -742.19975 + 785.62351 = 43.42376$$

$$C = 1 + \frac{1}{(3)(5)}\left(.05398 - \frac{1}{835}\right) = 1.00352$$

$$\frac{M}{C} = 43.27$$

With five degrees of freedom this value of chi-square leads us to reject the idea of a common λ. This is consistent with the previous conclusion but we must temper our conclusion because of the questionable way in which we obtained $\bar{x}$ and $\sum(1/x_i - 1/\bar{x})$.

In addition to formal chi-square goodness of fit tests, Lancaster plotted $-\log(1 - \hat{F}_T)$ versus T where T is the observed strike duration and $\hat{F}_T$ is the sample cumulative distribution function. The hazard function (or failure rate) $r(t)$ defined by $r(t) = f(t)/[1 - F(t)]$ in Section 9.5 is equal to the derivative of $-\log[1 - F(t)]$; that is

$$r(t) = \frac{d}{dt}\{-\log[1 - F(t)]\}.$$ (10.27)

The hazard function was shown to increase to a maximum and then to decline to a horizontal asymptote. Thus if the slope of $-\log[1 - \hat{F}(T)]$ appears to behave like the hazard function, we have additional evidence supporting the inverse Gaussian. This, in fact, was the case with the strike data.

10.6 A WORD FREQUENCY DISTRIBUTION

Several statistical studies in linguistics have used word frequency and sentence length for variables. Often, empirical distributions for these variables are used to develop suitable statistical models and to study important language characteristics of, for example, an author's specific work. In his extensive study on the topic, Yule (1944) investigated many linguistic data sets and proposed the use of the compound Poisson law for the distribution of word frequencies. Herdan (1956, 1960) arrived at the same conclusion based on a number of observed word frequency distributions.

For a specified text the observed word frequency distribution is obtained by counting the number of different words appearing once, twice, three times, and so on, in the text. The random variable is the word frequency of usage which is discrete. That it has a compound Poisson distribution can be shown briefly as follows.

Let r denote the number of times a specific word appears in a text written by an author who had a large vocabulary of distinct words available to write on a specific topic. Suppose the text consists of N words and suppose π is the long-term probability of occurrence for the word. Then the probability distribution of r can be approximated by the binomial model. Since N is expected to be large and π to be small, one can use the Poisson distribution in place of the binomial. Let $\theta = N\pi$. Then the probability distribution of r is

$$f(r; \theta) = \frac{e^{-\theta}\theta^r}{r!}, \qquad r = 0, 1, 2, \dots.$$

The author's vocabulary is expected to consist of a very large number of words, each having a small value of π. The parameter π can therefore be considered as a continuous variable; hence we can assume that the random variable θ has its own probability distribution $g(\theta)$. Thus the word frequency distribution for the text is a compound Poisson and is obtained by

$$f(r) = \int_0^\infty g(\theta) \frac{e^{-\theta}\theta^r}{r!} \, d\theta, \qquad r = 0, 1, 2, \ldots. \tag{10.28}$$

However, the word frequency of zero (i.e., $r = 0$) needs to be excluded from the distribution which starts at $r = 1$. Accordingly, the word frequency distribution is a truncated case of (10.28) and is given by

$$f_0(r) = [1 - f(0)]^{-1} \int_0^\infty g(\theta) \frac{e^{-\theta}\theta^r}{r!} \, d\theta, \qquad r = 1, 2, 3, \ldots,$$
$$\tag{10.29}$$

where $f(0)$ is obtained by substituting $r = 0$ in (10.28).

The choice of a distribution for $g(\theta)$ is crucial. A word frequency distribution is likely to be highly skewed to the right. It is fair to assume that a large number of different words will occur rarely, forming a high relative frequency on the left, and a very few words (an example being the word "the") will occur quite frequently, resulting in a very small relative frequency in the tail of the distribution. An appropriate choice for $g(\theta)$ is a distribution which can lead to a very skewed shape for the distribution in (10.29).

The generalized inverse Gaussian distribution discussed in Section 2.7 possesses the desired property stated above and was considered by Sichel (1975). He showed that this choice provided a satisfactory fit to almost all of the data sets he considered in his study.

Considering (2.25) as the density function of θ it can be shown that the frequency distribution in (10.28) simplifies to

$$f(r) = \frac{\mu^r K_{r+\gamma}(\lambda/\mu)(1 + 2\mu^2/\lambda)^{1/2}}{r!(1 + 2\mu^2/\lambda)^{(r+\gamma)/2} K_r(\lambda/\mu)}, \qquad r \geqslant 0, \tag{10.30}$$

where $K_r(\cdot)$ is a modified Bessel function of the third kind. Thus it follows that the word frequency distribution is given by

$$f_0(r) = A\left[\frac{\{\mu(1+2\mu^2/\lambda^{1/2}\}^r}{r!}\right] K_{r+\gamma}\left(\frac{\lambda}{\mu}\right)\left(1+\frac{2\mu^2}{\lambda}\right)^{1/2}, \qquad r \geqslant 1$$

(10.31)

where

$$A^{-1} = \left(1+\frac{2\mu^2}{\lambda}\right)^{\gamma/2} K_\gamma\left(\frac{\lambda}{\mu}\right) - K_\gamma\left[\left(\frac{\lambda}{\mu}\right)\left(1+\frac{2\mu^2}{\lambda}\right)^{1/2}\right].$$

Maximum likelihood estimation of all three parameters in (10.31) is tedious. No explicit expressions have been obtained for the moments of the distribution in (10.31) except for $E(r)$. Sichel (1975), using the recurrence relation,

$$f(r) = \left[1-\left(1+\frac{2\mu^2}{\lambda}\right)^{-1}\right]\left(\frac{r+\gamma-1}{r}\right)f(r-1)$$

$$+ \left(\frac{\lambda}{\mu}\right)^2\left[\left(1+\frac{2\mu^2}{\lambda}\right)^{1/2}-\left(1+\frac{2\mu^2}{\lambda}\right)^{-1/2}\right]^2 \frac{f(r-2)}{4r(r-1)},$$

(10.32)

gives explicit expressions for $f_0(1)$ and $f_0(2)$. He also obtains the first moment $E(r)$. Equating $f_0(1)$, $f_0(2)$, and $E(r)$ to their corresponding observed values, he then estimates μ, λ, and γ using the method of minimum chi-square. Since the expressions for $f_0(1)$, $f_0(2)$, and $E(r)$ are very complex he fitted this distribution to only 2 out of the 20 data sets that he analyzed. In the other cases the particular value of $\gamma = -\frac{1}{2}$ was considered. This value corresponds to the two-parameter inverse Gaussian and thus the word frequency distribution is simply a compound Poisson-inverse Gaussian. This latter distribution was also considered by Holla (1966) for his model of accident or disease proneness among individuals.

After substituting $\gamma = -\frac{1}{2}$ in (10.31) and simplifying the resulting expression, we get

$$f_0(r) = \frac{A}{r!}\left(\frac{\mu}{(1+2\mu^2/\lambda)^{1/2}}\right)^r K_{r-1/2}\left[\left(\frac{\lambda}{\mu}\right)\left(1+\frac{2\mu^2}{\lambda}\right)^{1/2}\right]$$

where

$$A^{-1} = \left[\pi \frac{\mu}{2\lambda} \left(1 + \frac{2\mu^2}{\lambda} \right)^{1/2} \right]^{1/2}$$

$$\times \left\{ \exp\left(-\frac{\lambda}{\mu} \right) - \exp\left[-\left(1 + \frac{2\mu^2}{\lambda} \right)^{1/2} \right] \right\}. \qquad (10.33)$$

The expression (10.33) for $f_0(r)$ can be written in a somewhat simpler form by letting

$$\frac{\lambda}{\mu} = \alpha$$

$$\frac{\lambda}{\mu} \left(1 + \frac{2\mu^2}{\lambda} \right)^{1/2} = \beta.$$

Then, substituting for μ and λ in (10.33), where

$$\mu = \frac{\beta^2 - \alpha^2}{2\alpha},$$

$$\lambda = \frac{\beta^2 - \alpha^2}{2},$$

we have

$$f_0(r) = \left(\frac{\pi}{\beta} \right)^{1/2} (e^{-\alpha} - e^{-\beta}) \left(\frac{\beta^2 - \alpha^2}{2\beta} \right)^r \frac{1}{r!} K_{r-1/2}(\beta), \qquad r \geqslant 1.$$

$$(10.34)$$

Again, both MLE and moment estimates are not explicitly known. Sichel (1975) considered $f(1)$ and $E(r)$ to obtain his estimates of parameters μ and λ and to compute the expected frequencies for the distribution in (10.34). He fitted 18 observed word frequency distributions to this two-parameter distribution function. Results of these 18 and two other fits for the three-parameter case are summarized in Table 2 of his paper. Since Sichel

Table 10.4   Statistics for 20 Word Frequency Distributions

Distribution	N	γ or $\hat{\gamma}$	$\hat{\mu}$	$\hat{\lambda}$	χ^2	df
1	2048	−.5	1.265	.109	10.63	15
2	1030	−.5	.489	.013	17.00	18
3	353	−.5	2.512	.157	5.91	12
4	1834	−.5	2.581	.431	14.48	23
5	5425	−.5	6.855	.283	43.10	48
6	3543	−.5	2.586	.239	27.31	33
7	2246	−.5	2.985	.178	25.42	29
8	965	−.5	.538	.030	16.89	14
9	1241	−.5	.194	.004	16.82	15
10	748	−.5	1.142	.070	16.33	15
11	548	−.5	.560	.026	11.87	11
12	593	−.5	.560	.018	8.60	13
13	1754	−.5	1.668	.133	20.82	26
14	529	−.5	4.526	.240	12.71	17
15	2648	−.5	2.772	.155	24.00	22
16	1345	−.5	1.515	.060	14.89	13
17	1906	−.5	1.891	.123	16.17	18
18	1073	−.5	1.769	.063	10.13	12
19	6001	−.920	12.979	.540	25.08	13*
20	8437	−.825	4.368	.061	17.19	13

*$P(\chi^2_{13} > 25.08) = .023$.

has used a slightly different reparameterization than ours, in Table 10.4 we give these results with estimates for μ, λ. We observe that all except distribution 19 are satisfactory fits using the compound Poisson model discussed here.

10.7   OTHER APPLICATIONS

10.7.1   Conversion Time for Convertible Bonds

Bachelier (1900) formulated a random walk model for securities and commodities. His model for price changes has come to be called

Brownian motion. Let $Z(t)$ be the price of a stock at the end of time period t. He assumed that differences $Z(t + T) - Z(t)$ were independent normal variables with mean zero and variance proportional to T.

Many modern writers (see Cootner, 19 4) have suggested that price data are better fitted by assuming $\ln Z(t + T) - \ln Z(t)$ gives independent Gaussian variables. That is, $\ln Z(t)$ is a Wiener process. When it is a Wiener process with positive drift, the first passage time to a barrier is, of course, an IG variable.

Under certain circumstances the time required to reach a certain level, or barrier, would be of interest. Whitmore (1976) mentions specifically the time required for a convertible bond to reach the call price for the first time.

10.7.2   Length of Employee Service

Whitmore (1979) uses a Wiener process with drift as a model of employee dissatisfaction (or, alternatively, satisfaction). He conceives of the idea of an underlying employee dissatisfaction at time t, $X(t)$, which is a normal random variable. It then seems reasonable to postulate the fluctuation of $X(t)$ with time and, finally, to assume that $X(t)$ is a Wiener process. Then an employee will leave the company when $X(t)$ reaches some barrier a for the first time. Thus the first passage time T can be thought of as length of employee service. However, unlike most of the cases considered in this work, the drift may be positive or negative. In either case the length of service is described by the usual inverse Gaussian distribution.

Whitmore (1979) fits the inverse Gaussian distribution to employee service times and finds that the distribution does a reasonably good job. The data sets are sets of grouped data and are fitted by both the methods of maximum likelihood and minimum chi-square. In most cases μ is estimated to be positive; in some cases negative. This is quite puzzling. Since the inverse Gaussian with both positive and negative drift v has mean $|(a - x_0)/v|$, there is no way to determine from the data whether v is positive or negative.

10.7.3   Net Maternity Function

In the study of a renewal process for the human population, Hadwiger (1940) basically applied the inverse Gaussian distribution to model what is known as the net maternity function in demography. He shows that the maternity function for the successive generations is the same as that of the first generation except for its parameters, which, in our notation for the inverse Gaussian, $IG(\mu, \lambda)$, are $n\mu$ and $n^2\lambda$, where n is the number of the successive generations (Keyfitz, 1977, p. 149). According to Jorgensen (1980), there seems to be no connection between the works of Tweedie and Hadwiger in the development of the inverse Gaussian distribution. In demography, the IG is known as the Hadwiger curve. Hoem (1976) has, perhaps for the first time, brought this application of the inverse Gaussian distribution to the attention of statisticians.

10.7.4   Wind Speed and Energy Evaluation

Bardsley (1980) suggested the use of the inverse Gaussian distribution to model wind speed and energy flux, particularly in situations where low speeds are rare. When the speeds are obtained as averages over long time intervals, it practically eliminates low speeds. Therefore, a modeling of the wind speed may need a threshold parameter in describing its distribution. For this reason, Stewart and Essenwanger (1978) fitted a three-parameter Weibull distribution and found it to be an improvement over the two-parameter Weibull fit. However, a positive value of the Weibull threshold parameter leads to the unrealistic condition of zero probability of wind speeds less than the threshold value. Moreover, an addition of third parameter introduces difficulties in parameter estimation. On the other hand, a two-parameter inverse Gaussian distribution can entail considerable displacement from zero, thus effecting a threshold value, while at the same time maintaining a significant amount of skewness (e.g., see Figure 2.1). This suggests that a three-parameter Weibull distribution could be replaced by a two-parameter inverse Gaussian distribution. Bardsley (1980) showed a close resemblance between the two distributions for some standardized pdf cases corresponding to the .05 and .1 contours of M, a measure of similarity defined by the maximum difference between the two distribution functions.

The wind energy flux is defined by

$$Z = \frac{X^3}{2\rho} \tag{10.35}$$

where X is the wind speed and ρ is the air density. Letting $X \sim \mathrm{IG}(\mu, \lambda)$, it follows from (2.5) that the mean energy flux is given by

$$\mu_z = \frac{(1 + 3\varphi^{-1} + 3\varphi^{-2})}{2\rho} \mu^3 \tag{10.36}$$

where $\varphi = \lambda/\mu$. Sometimes the only data available from a wind site are the averaged wind speeds over some relatively long time interval, say ΔT. However, it is desirable to estimate the mean wind energy flux over a much smaller time interval, say, Δt defined by

$$\Delta t = \frac{\Delta T}{K}$$

where K is an integer. Since the IG distribution possesses a reproductive property under summation, the properties of the wind speed distribution and energy flux for the time interval Δt can be derived easily from the given wind speed data corresponding to the longer time interval ΔT. If μ_T, φ_T are the IG parameters corresponding to ΔT, then the parameters for the Δt time interval case are

$$\mu_t = \mu_T$$

$$\varphi_t = \varphi_T \left(\frac{\Delta t}{\Delta T}\right).$$

Moreover, the mean wind energy flux is given by

$$\mu_z = \frac{[1 + 3\varphi_T^{-1}(\Delta T/\Delta t) + 3\varphi_T^{-2}(\Delta T/\Delta t)^{-2}]}{2\rho} \mu_T^3. \qquad (10.37)$$

The invariance property of the inverse Gaussian variable under a scale change suggests another application. If the wind speed increases with height in terms of the logarithmic or power law models, then a speed distribution at some higher level L_1 must be the same as that at a lower level L_0, except for a scale change. When the speed measurements are available at L_0, one can estimate the speed distribution at L_1, assuming the wind distribution is inverse Gaussian. The scale change is simply achieved by a change in mean μ, that is,

$$\mu_1 = a\mu_0$$

where a is a function of height.

10.7.5 Crop Field Size Distribution

In aerial surveys of agricultural land the crop field size distributions play a major role in the survey design. First, the resolution size for the sensor system should be decided so that the crop fields become identifiable from the acquired data. Second, a sampling design must take the field size and configuration into account to produce an efficient estimator with maximum economy in data. Based on such considerations Maxim and Harrington (1982) suggested the use of inverse Gaussian for modeling crop field sizes; also, because the field size distributions, in general, tend to be positively skewed. Their suggestion led Ferguson et al. (1986) to investigate the field size distributions for various crops in the United States and Canada. Field sizes were measured in length, width, and area for ten individual crops as well as for all crops combined in a state. Table 10.5 shows the location (state), crop type, and number of fields for which the field size data were available. These data were used to

Table 10.5   Numbers of Fields of the Most Prevalent Crops, and of All Crops, in Each State

State	Crop type	No. of fields	State	Crop type	No. of fields
Canada	Idle	1315	Missouri	Soybeans	980
	Spring wheat	1126		All crops	3009
	All crops	3337	Montana	Idle	1441
Illinois	Corn	2051		Spring wheat	726
	Soybeans	2072		Winter wheat	899
	All crops	6388	Nebraska	Alfalfa	1189
Indiana	Corn	3358		Corn	1386
	Hay	792		Sorghum	952
	Soybeans	3219		Winter wheat	1503
		944		All crops	8770
	All crops	10,962			

State	Crop	Value		State	Crop	Value
Iowa	Alfalfa	1674		N. Dakota	Barley	889
	Corn	5451			Oats	853
	Hay	1204			Spring wheat	1503
	Oats	1624			All crops	11,835
	Soybeans	4639				
	All crops	17,381		Oklahoma	Winter wheat	697
					All crops	2221
Kansas	Sorghum	1620				
	Winter wheat	2557		S. Dakota	Alfalfa	1043
	All crops	9683			Corn	1071
					Oats	1108
Minnesota	Corn	978			Spring wheat	706
	Oats	731			All crops	6886
	Spring wheat	748				
	All crops	5611		Texas	Winter wheat	800
					All crops	4428
Mississippi	Soybeans	938				
	All crops	2013		Total		56,408 97,456

obtain inverse Gaussian fits for each crop and each state as well as for all crops in a state or all states for a crop type. The goodness of fit and other field size evaluation results are presented in details in Ferguson et al. (1986). This empirical study showed that the inverse Gaussian distribution can be used as a model for crop field size when measured in terms of area and width, but not length.

10.7.6   Risk Analysis

Berg (1980) proposed the use of the inverse Gaussian in risk analysis. He attributed this application to Seal (1969, 1978) and showed that both the inverse Gaussian and the lognormal distributions provide tractable loglinear models for the claim cost distribution. Using a multidimensional loglinear parameterization, he obtained the maximum likelihood estimates of the parameters and discussed the asymptotic efficiency.

11

Additional Topics

11.1 BIVARIATE INVERSE GAUSSIAN DISTRIBUTIONS

With the many results analogous to those for the normal, it seems that a bivariate distribution should exist with properties similar to the bivariate normal. However, attempts to formulate such a bivariate inverse Gaussian point out once more the very special nature of the normal distribution. The beautiful properties enjoyed by the bivariate normal distribution are apparently lacking from other bivariate distributions.

What would we require of a distribution for it to be called a bivariate inverse Gaussian distribution? We would like a unimodal function of x and y, skewed to the right in both x and y with both marginal distributions and both conditional distributions inverse Gaussian. Failing to have both marginals and both conditionals inverse Gaussian, we would like to have at least one of the marginal distributions inverse Gaussian and the conditional distribution of the other variable inverse Gaussian.

Although it is attractive for such a distribution to arise from some process like the Wiener process, we do not regard this as crucial. The univariate inverse Gaussian gives a good description of many skewed populations in cases where it is not possible to envision a Wiener process. Rather the distribution is used because it seems to fit the data. Thus if we could develop a bivariate inverse Gaussian which seemed to fit populations skewed in both variables, this would justify its use upon this utility.

We now present three different approaches to construction of a bivariate inverse Gaussian distribution.

Case 1

Wasan (1968) gave a bivariate inverse Gaussian distribution which arises quite naturally as the joint distribution of first passage times to a first and second barrier. Consider a Wiener process $X(t)$ beginning with $X(0) = x_0$ with positive drift v and variance σ^2. Choose a and b so that $x_0 < a < b$ and consider the first passage time T_1 from x_0 to a and T_2 from a to b. Then T_1 and T_2 are independent inverse Gaussian variables with parameters $\mu_1 = (a - x_0)/v$, $\lambda_1 = (a - x_0)^2/\sigma^2$ and $\mu_2 = (b - a)/v$, $\lambda_2 = (b - a)^2/\sigma^2$. Consider now the joint distribution of T_1 and $T_3 = T_1 + T_2$. It is given by the density function

$$f(t_1, t_3) = \frac{1}{2\pi}\left[\frac{\lambda_1\lambda_2}{t_1^3(t_3 - t_1)^3}\right]^{1/2}$$
$$\times \exp\left\{-\frac{\lambda_1(t_1 - \mu_1)^2}{2\mu_1^2 t_1} - \frac{\lambda_2(t_3 - t_1 - \mu_2)^2}{2\mu_2^2(t_3 - t_1)}\right\} \quad (11.1)$$

with $0 < t_1 < t_3 < \infty$. Marginally, T_1 is an inverse Gaussian variable. Conditionally on $T_1 = t_1$, T_3 is a translated inverse Gaussian variable; that is, $t_3 = t_1 + \mathrm{IG}(\mu_2, \lambda_2)$. Because for the first passage times $\lambda_i/\mu_i^2 = v^2/\sigma^2 = \text{constant}$, $T_3 = T_1 + T_2$ is also an inverse Gaussian variable. That is,

$$T_3 \sim \mathrm{IG}[\mu_1 + \mu_2, v^2(\mu_1 + \mu_2)^2/\sigma^2]. \quad (11.2)$$

Since $\mu_1 + \mu_2 = (b - x_0)/v$

$$T_3 \sim IG[(b - x_0)/v, (b - x_0)^2/\sigma^2]. \tag{11.3}$$

The last observation also follows directly from the realization that T_3 is the first passage time from x_0 to b. In summary, T_1 and T_3 are marginally distributed as inverse Gaussian, and conditional on $T_1 = t_1$, T_3 is a translated inverse Gaussian variable. It remains as a minor puzzle to find a simple expression for the conditional density of T_1 given $T_3 = t_3$.

If we considered a bivariate distribution with the form given in equation (11.1) without the stipulation of a common origin from a single Wiener process, we could no longer claim that $\lambda_i/\mu_i^2 = $ constant, $i = 1, 2$. Without this condition, T_3 is no longer an inverse Gaussian variable but the conditional distribution is still that of a translated inverse Gaussian variable.

With or without the condition that $\lambda/\mu^2 = $ constant, any probability considerations involving T_1 and T_3 can be resolved by using the distribution of T_1 and $T_3 - T_1$, independent IG variables. So a bivariate distribution with the form in equation (11.1) seems artificial and not very satisfying.

Case 2

Al-Hussaini and Abd-El-Hakim (1981) gave a bivariate inverse Gaussian distribution which they constructed from given marginals. If $f_1(x_1)$ and $f_2(x_2)$ are the pdf of random variables X_1 and X_2, respectively, then a bivariate pdf $f_{12}(x_1, x_2)$ may be constructed as follows:

$$f_{12}(x_1, x_2) = f_1(x_1)f_2(x_2)[1 + K(x_1 - a_1)$$
$$\times (x_2 - a_2)h_1(x_1)h_2(x_2)] \tag{11.4}$$

where $h_i(x_i)$, $i = 1, 2$ are chosen so that $f_{12}(x_1, x_2)$ is a pdf, with

$$a_i = \frac{E[X_i h_i(X_i)]}{E[h_i(X_i)]} \tag{11.5}$$

and K as a constant satisfying

$$1 + K(x_1 - a_1)(x_2 - a_2)h_1(x_1)h_2(x_2) \geqslant 0 \qquad (11.6)$$

for all (x_1, x_2) in the domain of definition of $f_{12}(x_1, x_2)$. It is obvious that $f_{12}(x_1, x_2)$ yields $f_1(x_1)$ and $f_2(x_2)$ as marginal pdf's.
Letting $X_i \sim \text{IG}(\mu_i, \lambda_i)$, and

$$h_i(x_i) = \exp\left(-\frac{\lambda_i(x_i - \mu_i)^2}{2\mu_i^2 x_i}\right), \qquad (11.7)$$

it follows that $a_i = \mu_i$, $i = 1, 2$ and

$$f_{12}(x_1, x_2) = x[1 + \rho\Psi(x_1, x_2)]\frac{1}{2\pi}\left(\frac{\lambda_1\lambda_2}{x_1^3 x_2^3}\right)^{1/2}$$

$$\times \exp\left\{-\left[\frac{\lambda_1(x_1 - \mu_1)^2}{2\mu_1^2 x_1} + \frac{\lambda_2(x_2 - \mu_2)^2}{2\mu_2^2 x_2}\right]\right\}$$

$$(11.8)$$

$$\mu_i > 0, \quad \lambda_i > 0, \quad x_i > 0 \qquad (i = 1, 2),$$

and $f_{12}(x_1 x_2) = 0$ outside the first quadrant of the plane, where

$$\Psi(x_1, x_2) = 8\sqrt{\frac{\lambda_1\lambda_2}{\mu_1^3\mu_2^3}}(x_1 - \mu_1)(x_2 - \mu_2)$$

$$\times \exp\left\{-\left[\frac{\lambda_1(x_1 - \mu_1)^2}{2\mu_1^2 x_1} + \frac{\lambda_2(x_2 - \mu_2)^2}{2\mu_2^2 x_2}\right]\right\}.$$

$$(11.9)$$

It follows from (11.6) that the correlation coefficient ρ between X_1 and X_2 is bounded by $-a$ and b, where

$$-a = \max\left(-1, -\frac{1}{M}\right), \quad b = \min\left(\frac{1}{N}, 1\right)$$

with

$$M = \max_{x_1, x_2, > 0} \Psi(x_1, x_2), \quad -N = \min_{x_1, x_2 > 0} \Psi(x_1, x_2).$$

It is obvious from (11.8) that X_1 and X_2 are independent if and only if they are uncorrelated.

The bivariate distribution function corresponding to (11.8) is given by

$$F_{12}(x_1, x_2) = F_1(x_1)F_2(x_2) + \rho H(x_1, x_2) \tag{11.10}$$

where $F_i(x_i)$ is the univariate IG distribution function [e.g., refer to Equation (2.14) for an exact expression of F], $i = 1, 2$, and

$$H(x_1, x_2) = \int_0^{x_1} \int_0^{x_2} \Psi(u_1, u_2) f_1(u_1) f_2(u_2) \, du_1 \, du_2$$

$$= 16 \sqrt{\frac{\lambda_1 \lambda_2}{\mu_1 \mu_2}} \exp\left[4\left(\frac{\lambda_1}{\mu_1} + \frac{\lambda_2}{\mu_2}\right)\right] \Phi(-C_1)\Phi(-C_2)$$

$$\tag{11.11}$$

where Φ denotes the cdf for the standard normal and

$$C_i = \left[2\left(\frac{4\lambda_i}{\mu_i} + y_i^2\right)\right]^{1/2}$$

$$y_i = \frac{\sqrt{\lambda_i}(x_i - \mu_i)}{\mu_i \sqrt{x_i}}$$

$$i = 1, 2. \tag{11.12}$$

The expressions in (11.10) to (11.12) can be obtained by making use of the transformation considered in Theorem 2.1.

Denote the bivariate IG distributed pair of random variables as $(X_1, X_2) \sim IG(\mu_1, \mu_2, \lambda_1, \lambda_2, \rho)$. Several of the univariate IG properties discussed in Chapter 2 can be easily shown to hold for the bivariate IG. For example, if $Z_i = a_i X_i$, $(a_i > 0)$, $i = 1, 2$, and

$(X_1, X_2) \sim \text{IG}(\mu_1, \mu_2, \lambda_1, \lambda_2, \rho)$, then $(Z_1, Z_2) \sim \text{IG}(a_1\mu_1, a_2\mu_2, a_1\lambda_1, a_2\lambda_2, \rho)$. For further discussion refer to Al-Hussaini and Abd-El-Hakim (1981).

Although the marginal distributions of X_1 and X_2 are inverse Gaussian, their conditional distributions are not. Considering the conditional pdf of X_1 given $X_2 = x_2$, it follows that

$$h(x_1 \mid x_2) = \frac{f_{12}(x_1, x_2)}{f_2(x_2)}$$

$$= f_1(x_1)[1 + \rho\Psi(x_1, x_2)]. \tag{11.13}$$

Next, the conditional mean of X_1 given $X_2 = x_2$ is obtained as

$$E[X_1 \mid x_2] = \mu_1 + 8\rho \sqrt{\frac{\lambda_1\lambda_2}{\mu_1^3\mu_2^3}}(x_2 - \mu_2)$$

$$\times \exp\left[-\frac{\lambda_2(x_2 - \mu_2)^2}{2\mu_2^2 x_2}\right]$$

$$\times \frac{1}{\sqrt{2}}\int_0^\infty x_1(x_1 - \mu_1)\sqrt{\frac{2\lambda_1}{2\pi x_1^3}}$$

$$\times \exp\left[-\frac{\lambda_1(x_1 - \mu_1)^2}{2\mu_1^2 x_1}\right]dx_1,$$

where the integral part is simply equal to $\text{Var}(X_1)$ with λ_1 replaced by $2\lambda_1$, that is, $\mu_1^3/2\lambda_1$. Hence,

$$E[X_1 \mid x_2] = \mu_1 + \rho\frac{\sigma_1}{\sigma_2}(x_2 - \mu_2)2\sqrt{2}$$

$$\times \exp\left[-\frac{\lambda_2(x_2 - \mu_2)^2}{2\mu_2^2 x_2}\right], \tag{11.14}$$

where

$$\sigma_i^2 = \frac{\mu_i^3}{\lambda_i}, \qquad i = 1, 2.$$

By symmetry, one can write the conditional pdf and mean of X_2 given $X_1 = x_1$.

Note that the expression in (11.14) has some analogy with the normal, though it is not linear with respect to x_2 as is the case with the normal distribution. Furthermore,

(i) $E[X_1 | \mu_2] = \mu_1$

(ii) $E[X_1 | x_2] \to \mu_1$ as $x_2 \to \infty$.

Its overall behavior depends upon the parametric values of ρ, μ_1 and λ_1 and can further be investigated by equating to zero the partial derivative of the conditional mean with respect to x_2 and finding the critical points where it changes the direction.

Case 3

Another bivariate IG distribution can be constructed using what Mardia (1970) called the trivariate reduction. This approach involves starting with three independent random variables, say X_1, X_2, X_3 with corresponding cdf's $F(x_i, \theta_i)$, $i = 1, 2, 3$, and finding a function T such that the cdf of $X = T(X_1, X_2)$ is $F(x; \theta_1 + \theta_2)$. Then X and another variable $Y = T(X_1, X_3)$ has a bivariate distribution. This is possible for IG when the set of independent variables are

$$X_i \sim \text{IG}(k_i \mu, k_i^2 \lambda), \qquad i = 1, 2, 3$$

and the pair of variables (X, Y) are defined by

$$X = X_1 + X_2, \qquad Y = X_1 + X_3.$$

Clearly, X and Y have IG marginals.

This approach has a drawback in the sense that the correlation coefficient ρ between X and Y is never negative. This is because

$$\rho = \frac{\sigma_1^2}{[(\sigma_1^2 + \sigma_2^2)(\sigma_1^2 + \sigma_3^2)]^{1/2}},$$

where $\sigma_i^2 = \text{Var}(X_i)$, $i = 1, 2, 3$.

11.2 MULTIVARIATE INVERSE GAUSSIAN DISTRIBUTIONS

The formulation given by Wasan for the bivariate distribution as discussed in case 1 of the last section was also extended by Wasan to the multivariate case. Consider $x_0 < x_1 \cdots < x_p$ and consider the first passage times $T_1, T_2, T_3, \ldots$, from x_0 to x_1, x_1 to $x_2, \ldots, x_{p-1}$ to x_p of a Weiner process with positive drift v and variance σ^2. Then

$$T_i \text{ indep IG}[(x_i - x_{i-1})/v, (x_i - x_{i-1})^2/\sigma^2]. \tag{11.15}$$

Note that $\lambda_i/\mu_i^2 = v/\sigma^2$, a constant, so that the sum of any set of T_i's is also inverse Gaussian. In particular,

$$Y_1 = T_1 \sim \text{IG}$$
$$Y_2 = T_1 + T_2 \sim \text{IG}$$
$$\vdots$$
$$Y_p = T_1 + T_2 + \cdots + T_p \sim \text{IG}. \tag{11.16}$$

The joint distribution of the Y's is given by the density function

$$f(y_1, y_2, \ldots, y_p) = (2\pi)^{-p/2} \left[\frac{\lambda_1 \lambda_2 \cdots \lambda_p}{y_1^3 (y_2 - y_1)^3 \cdots (y_p - y_{p-1})^3} \right]^{1/2}$$
$$\times \exp \left[-\frac{\lambda_1(y_1 - \mu_1)^2}{2\mu_1^2 y_1} - \frac{\lambda_2(y_2 - y_1 - \mu_2)^2}{2\mu_2^2(y_2 - y_1)} \right.$$
$$\left. - \cdots - \frac{\lambda_p(y_p - y_{p-1} - \mu_p)^2}{2\mu_p^2(y_p - y_{p-1})} \right] \tag{11.17}$$

where $0 < y_1 < y_2 < \cdots < y_p$. Now the distribution of any set of consecutive Y's conditional on the preceding Y's is easily expressed as a set of translated IG variables; however, the conditional distribution of a set of Y's given succeeding y's is not so easily expressed. A multivariate distribution having the form in equation (11.6) can also be given for arbitrary λ_i and μ_i. In this case, however, the marginal distributions would not be inverse Gaussian.

As in the above case, analogous multivariate distributions to the bivariate distributions given in cases 2 and 3 of the previous section can also be developed.

11.3 AN INVERSE GAUSSIAN PROCESS

In Chapter 3 we obtained the inverse Gaussian as the distribution of first passage time in a Wiener process. Sometimes the process is opposite to the Wiener case where the time duration acts as a process variable and the state as a time variable. For example, this may arise in monitoring the controls on money supply during an inflationary period or in studying the time of first overload of an electric circuit. When time duration is assumed distributed as inverse Gaussian with certain mean and variance, one can define a new process in a manner similar to the Wiener process. This new process has been known as an inverse Gaussian process (Wasan, 1968) or the first passage time process with Brownian motion (Basu and Wasan, 1974).

As in Wasan (1968), we now define an inverse Gaussian process with mean increment δ and scale parameter η as a stochastic process $\{X(w); w \geqslant 0\}$ with the following properties:

(i) $X(w)$ has independent increments; for every pair of disjoint intervals (w_1, w_2), (w_2, w_3) with $w_1 < w_2 < w_3 < w_4$, the random variables $X(w_2) - X(w_1)$ and $X(w_4) - x(w_3)$ are independent.

(ii) Each increment $X(w + z) - X(z)$ has an inverse Gaussian distribution with mean δw and scale parameter ηw^2.

(iii) $X(0) = 0$ with probability one.

Suppose that the process $X(w)$ started at $x_0 = X(w_0)$ and $w > w_0$. Then the conditional probability,

$$P[X(w) < x \mid X(w_0) = x_0]$$

$$= P[X(w) - X(w_0) \leqslant x - x_0]$$

$$= (w - w_0)\left(\frac{\lambda}{2\pi}\right)^{1/2} \int_0^{x - x_0} y^{-3/2}$$

$$\times \exp\left(-\frac{\eta[y - \delta(w - w_0)^2]}{2\delta^2 y}\right) dy. \tag{11.18}$$

Thus the density function of $X(w)$ given that $X(w_0) = x_0$ is obtained from (11.18) as

$$
f(x; x_0, w, w_0) = \frac{(w - w_0)\lambda^{1/2}}{[2\pi(x - x_0)^3]^{1/2}}
$$

$$
\times \exp\left(-\frac{\eta[x - x_0) - \delta(w - w_0)]^2}{2\delta^2(x - x_0)}\right),
$$

$$
x > x_0. \tag{11.19}
$$

From (11.19),

$$
E[X(w) - X(w_0)] = \delta(w - w_0)
$$

$$
\text{Var}[X(w) - X(w_0)] = \frac{\delta^3(w - w_0)}{\eta}.
$$

If $w_1 < w_2 < w_3$, then $X(w_3) - X(w_1)$ can be written as the sum of two independent inverse Gaussian variables $[X(w_3) - X(w_2)]$ and $[X(w_2) - X(w_1)]$ with means $\delta(w_3 - w_2)$, $\delta(w_2 - w_1)$ and scale parameters $\eta(w_3 - w_2)^2$, $\eta(w_2 - w_1)^2$, respectively. These random variables satisfy the additive property of IG given in Chapter 2. Thus the sum $X(w_3) - X(w_1)$ has inverse Gaussian distribution with mean $\delta(w_3 - w_1)$ and variance $\delta^3(w_3 - w_1)/\eta$. Accordingly, property (i) of an IG process as defined here is consistent with its property (ii). Those further interested in the mathematics of an IG process should refer to Basu and Wasan (1974), who have formulated it and studied its many properties in a rigorous manner.

Bibliography

Ahsanullah, M., and Kirmani, S. N. U. A. (1984). A characterization of the Wald distribution. *Naval Res. Logist. Quart., 31*: 155–158.

Aitchison, J., and Dunsmore, I. R. (1975). *Statistical Prediction Analysis,* Cambridge: Cambridge University Press.

Aitkinson, A. C. (1979). The simulation of generalized inverse Gaussian, generalized hyperbolic, gamma and related random variables. Research Report No. 52, Dept. Theor. Statist., Aarhus University, Aarhus, Denmark.

Al-Hussaini, E. K., and Abd-El-Hakim, N. S. (1980). Bivariate inverse Gaussian distribution. *Egypt. Statist. J., 24*(1):100–117.

Al-Hussaini, E. K., and Abd-El-Hakim, N. S. (1981). Bivariate inverse Gaussian distribution. *Ann. Inst. Statist. Math., 33*:57–66.

Amoh, R. K. (1984). Estimation of parameters in mixtures of inverse Gaussian distributions. *Commun. Statist. Theor. Meth., 13*:1031–1043.

Amoh, R. K. (1985). Estimation of a discriminant function from a mixture of two inverse Gaussian distributions when sample size is small. *J. Statist. Comput. Simul., 20*:275–286.

Athreya, K. B. (1986). Another conjugate family for the normal distribution. *Statist. Probab. Lett.*, *4*:61–64.

Bachelier, L. (1900). Theorie de la speculation. *Ann. Sci. Ec. Norm. Super.*, Paris, *17*(3):21–86.

Banerjee, A. K. (1977). Bivariate inverse Gaussian distribution (Abstract). *Bull. Inst. Math. Statist.*, *6*:138.

Banerjee, A. K., and Bhattacharyya, G. K. (1976). A purchase incidence model with inverse Gaussian interpurchase times. *J. Amer. Statist. Ass.*, *71*:823–829.

Banerjee, A. K., and Bhattacharyya, G. K. (1979). Bayesian results for the inverse Gaussian distribution with an application. *Technometrics*, *21*:247–251.

Bardsley, W. E. (1980). Note on the use of the inverse Gaussian distribution for wind energy application. *J. Appl. Meteorol.*, *19*:1126–1130.

Barlow, R. E., and Proschan, F. (1965). *Mathematical Theory of Reliability*. New York: Wiley.

Barndorff-Nielson, O. (1978a). *Information and Exponential Families*. Chichester: Wiley.

Barndorff-Nielsen, O. (1978b). Hyperbolic distributions and distributions on hyperbolae. *Scand. J. Statist.* *5*:151–159.

Barndorff-Nielsen, O., and Blaesild, P. (1980). Hyperbolic distributions. *Encyclopedia of Statistical Sciences*. New York: Wiley.

Barndorff-Nielsen, O., and Blaesild, P. (1981). Hyperbolic distributions and ramifications: contributions to theory and application, in *Statistical Distributions in Scientific Work*, Vol. 4, Taillie et al., eds. Dordrecht, Holland: Reidel.

Barndorff-Nielsen, O., and Blaesild, P. (1983a). Exponential models with affine dual foliations. *Ann. Statist.*, *11*:753–769.

Barndorff-Nielsen, O., and Blaesild, P. (1983b). Reproductive exponential families, *Ann. Statist.*, *11*:770–782.

Barndorff-Nielsen, O., and Halgren, C. (1977). Infinite divisibility of the hyperbolic and generalized inverse Gaussian distributions. *Z. Wahrscheinlichkeitstheorie Verw. Geb.*, *38*:309–312.

Barndorff-Nielsen, O., Blaesild, P., and Halgren, C. (1978). First hitting time models for the generalized inverse Gaussian distribution. *Stoch. Processes Appl.*, 7:49–54.

Barndorff-Nielsen, O., Blaesild, P., Jensen, J. L., and Jorgensen, B. (1982). Exponential transformation models. *Proc. Roy. Soc. Lond.*, *A379*:41–65.

Bartlett, M. S. (1937). Properties of sufficiency and statistical tests. *Proc. Roy. Soc. Lond.*, *A160*:268–282.

Bartlett, M. S. (1966). *An Introduction to Stochastic Processes.* London: Cambridge University Press.

Basu, A. K., and Wasan, M. T. (1974). On the first passage time process of Brownian motion with positive drift. *Scand. Actuar. J.*, 144–150.

Berg, P. T. (1980). Two pragmatic approaches to loglinear claim cost analysis. *Astin Bull.*, *11*:77–90.

Bhattacharyya, G. K. (1982). Fatigue failure models—Birnbaum–Saunders vs. inverse Gaussian. *IEEE Trans. Reliab.* R-31 (5):439–440.

Bhattacharyya, G. K., and Fries, A. (1982). Inverse Gaussian regression and accelerated life tests, in *Survival Analysis*, edited by John Crowley and Richard A. Johnson, IMS Lecture Notes, Monograph Series, pp. 101–118.

Boag, J. W. (1949). Maximum likelihood estimates of the proportion of patients cured by cancer therapy. *J. Roy. Statist. Soc. B11*:15–44.

Bogaard, J. M. (1979). Interpreting of indicator dilution curves using a random model. Thesis, Medical Faculty, University of Rotterdam, Netherlands.

Bondi, A. B. (1977). Stochastic models of single neuronal spike trains. M.Sc. Dissertation, University College, London.

Box, G. E. P. (1980). *J. Roy. Statist. Soc. A143*: 383–430.

Box, G. E. P., and Cox, D. R. (1964). An analysis of transformations. *J. Roy. Stat. Soc.*, *B26*:211.

Box, George E. P., and Tiao, George C. (1973). *Bayesian Inference in Statistical Analysis.* Reading, Mass.: Addison-Wesley.

Brown, R. (1828). A brief account of microscopial observations made in the months of June, July, and August, 1827, on the particles contained in the pollen of plants; and on the general existence of active molecules in organic and inorganic bodies. *Phil. Mag.*, Series 2, *4*:161–173.

Brush, S. G. (1968). A history of random processes. *Arch. Hist. Exact Sci.*, *5*:1–36. Reprinted in *Studies in the History of Statistics and Probability*, Vol. II, edited by M. G. Kendall and R. L. Plackett, New York: Macmillan, 1977.

Capocelli, R. M., and Ricciardi, L. M. (1972). On the inverse of the first passage time probability problem. *J. Appl. Probab.*, *9*:270–287.

Chan, M. Y., Cohen, A. C., and Whitten, B. J. (1983). The standardized inverse Gaussian distribution—Tables of the cumulative probability function. *Commun. Statist.*, *B12*:423–442.

Chan, M. Y., Cohen, A. C., and Whitten, B. J. (1984). Modified maximum likelihood and modified moment estimators for the three-parameter inverse Gaussian distribution. *Commun. Statist. Simul. Comp.*, *13*:47–68.

Cheng, R. C. H., and Amin, N. A. K. (1981). Maximum likelihood estimation of parameters in the inverse Gaussian distribution, with unknown origin. *Technometrics*, *23*:257–263.

Chhikara, R. S. (1972). Statistical inference related to the inverse Gaussian distribution. Ph.D. Dissertation, Oklahoma State University, Stillwater.

Chhikara, R. S. (1975). Optimum tests for the comparison of two inverse Gaussian distribution means. *Austral. J. Statist.*, *17*:77–83.

Chhikara, R. S., and Folks, J. L. (1974). Estimation of the inverse Gaussian distribution function. *J. Amer. Statist. Ass.*, *69*:250–254.

Chhikara, R. S., and Folks, J. L. (1975). Statistical distributions related to the inverse Gaussian. *Commun. Statist.*, *4*:1081–1091.

Chhikara, R. S., and Folks, J. L. (1976). Optimum tests procedures for the mean of first passage time in Brownian motion with positive drift (inverse Gaussian distribution). *Technometrics*, *18*:189–193.

Chhikara, R. S., and Folks, J. L. (1977). The inverse Gaussian distribution as a lifetime model. *Technometrics*, *19*:461–468.

Chhikara, R. S., and Guttman, Irwin (1982). Prediction limits for the inverse Gaussian distribution. *Technometrics*, *24*:319–324.

Cox, D. R., and Miller, H. D. (1965). *The Theory of Stochastic Processes.* London: Methuen.

Cootner, P., ed. (1964). *The Random Character of Stock Market Prices,* Cambridge, Mass.: M.I.T. Press.

Cramer, H. (1946). *Mathematical Methods of Statistics.* Princeton, N.J.: Princeton University Press.

Cressie, N., Davis, A. S., Folks, J. L., and Policello II, G. E. (1981). The moment-generating function and negative integer moments. *Amer. Statist.,* 35:148–150.

Davis, A. S. (1977). Linear statistical inference as related to the inverse Gaussian distribution. Ph.D. Dissertation, Oklahoma State University, Stillwater.

Davis, A. S. (1980). Use of the likelihood ratio test on the inverse Gaussian distribution. *Amer. Statist.,* 34:108–110.

Desmond, A. F. (1986). On the relationship between two fatigue-life models. *IEEE Trans. Reliab.,* R-35(2):167–169.

Doob, J. L. (1953). *Stochastic Processes.* New York: Wiley.

Easterling, R. G. (1976). Goodness of fit and parameter estimation. *Technometrics,* 18:1–9.

Einstein, A. (1905). Uber die von der molekularkinetischen Theorie der Warme geforderte Bewegung von in ruhenden Flussigkeiten suspendierten Teilchen. *Ann. Phys.,* 17:549–560.

Einstein, A. (1905). Investigations on the theory of the Brownian movement, edited with notes by R. Furth, translated by A. D. Cowper, 1956 ed. New York: Dover.

Erdely, A., ed. (1953). *Higher Transcendental Functions,* Vol. I. New York: McGraw-Hill.

Farewell, V. T., and Prentice, R. L. (1977). A study of distributional shape in life testing. *Technometrics,* 19:69–75.

Federighi, E. T. (1959). Extended tables of the percentage points of Student's *t* distribution. *J. Amer. Statist. Ass.,* 54:683–688.

Feller, W. (1957). *An Introduction to Probability Theory and Its Applications,* Vol. I. New York: Wiley.

Feller, W. (1966). *An Introduction to Probability Theory and Its Applications*, Vol. II. New York: Wiley.

Ferger, W. F. (1931). The nature and use of the harmonic mean. *J. Amer. Statist. Ass.*, *26*:36–40.

Ferguson, M., Badhwar, G., Chhikara, R., and Pitts, D. (1986). Field size distributions for selected agricultural crops in the United States and Canada, *Remote Sensing Environ.*, *19*:25–45.

Fienberg, S. E. (1974). Stochastic models for single neuron firing trains: a survey. *Biometrics*, *30*:399–427.

Fisher, R. A., Thornton, H. G., and Mackenzie, W. A. (1922). The accuracy of the plating method estimating the density of bacterial populations. *Ann. Appl. Biol.*, *9*:325–359.

Fisher, W. J. (1932). Harvard College Observatory Circular No. 375, Cambridge, Mass.

Folks, J. L., and R. S. Chhikara (1978). The inverse Gaussian distribution and its statistical application—a review. *J. Roy. Stat. Soc.*, *B40*:263–289.

Folks, J. L., Pierce, D. A., and Stewart, C. (1965). Estimating the fraction of acceptable product. *Technometrics*, *7*:43–50.

Fries, A., and Bhattacharyya, G. K. (1983). Analysis of two-factor experiments under an inverse Gaussian model. *J. Amer. Statist. Ass.*, *78*:820–826.

Gacula, M. C., Jr., and Kubala, J. J. (1975). Statistical models for shelf life failures. *J. Food Sci.*, *40*:404–409.

Gani, J., and Prabhu, N. U. (1963). A storage model with continuous infinitely divisible inputs. *Proc. Camb. Phil. Soc.*, *59*:417–430.

Gerstein, G. L., and Mandelbrot, B. (1964). Random walk models for the spike activity of a single neuron. *Biophys. J.*, *4*:41–68.

Gertsbakh, I. B., and Kordonskiy, K. B. (1969). *Models of Failure*. New York: Springer-Verlag.

Gnedenko, B. V., and Kolmogorov, A. N. (1954). *Limit Distributions for Sums of Independent Random Variables*. Reading, Mass.: Addison-Wesley.

Good, I. J. (1953). The population frequencies of species and the estimation of population parameters. *Biometrika*, *40*:237–264.

Gray, H. L., and Schucany, W. R. (1972). *The Generalized Jackknife Statistic.* New York: Marcel Dekker.

Gupta, S. S., and Yang, H. (1982). On subset selection procedure for inverse Gaussian populations. Technical Report, Purdue University, Lafayette, Ind.

Guttman, Irwin (1970). *Statistical Tolerance Regions: Classical and Bayesian,* London: Charles Griffin.

Hadwiger, H. (1940). Eine analytische Reproducktions—funktion fur biologische Gesamtheiten. *Skand. Aktuarietidskr., 23*:101–113.

Haldane, J. B. S. (1945). A labour-saving method of sampling. *Nature, 155*:49–50.

Halgren, C. (1979). Self-decomposability of the generalized inverse Gaussian and hyperbolic distributions. *Zeit. Wahrscheinlichkeitsth, 47*:13–17.

Hasofer, A. M. (1964). A dam with inverse Gaussian input. *Proc. Camb. Phil. Soc., 60*:931–933.

Haybittle, J. L. (1959). The estimation of the proportion of patients cured after treatment for cancer of the breast. *Brit. J. Radiol., 32*:725–733.

Herdan, G. (1956). *Language as Choice and Change.* Groningen: Noordhoff.

Herdan, G. (1960). *Type-Token Mathematics: A Textbook of Mathematical Linguistics.* The Hague: Mouton.

Hoem, J. M. (1976). The statistical theory of demographic rates: a review of current developments (with discussion). *Scand. J. Statist., 3*:169–185.

Holden, A. V. (1976). Models of the stochastic activity of neurons. *Lecture Notes in Biomathematics,* No. 12, Berlin: Springer-Verlag.

Holla, M. S. (1966). On a Poisson-inverse Gaussian distribution. *Metrika, No. 11*: 115–121.

Huff, B. W. (1974). A comparison of sample path properties for the inverse Gaussian and Bessel processes. *Scand. Actuarial J.,* 157–166.

Huff, B. W. (1975). The inverse Gaussian distribution and Root's barrier construction. *Sankhyā, 37A*:345–353.

Iwase, K., and Seto, N. (1983). Uniformly minimum variance unbiased estimation for the inverse Gaussian distribution. *J. Amer. Statist. Assoc., 78*:660–663.

Jain, G. C., and Khan, M. S. H. (1979). On an exponential family. *Math. Opf. Statist.*, *10*:153–168.

Johnson, N. L., and Kotz, S. (1970). *Distributions in Statistics: Continuous Univariate Distributions 1.* Boston: Houghton Mifflin.

Jones, G., and Cheng, R. C. H. (1984). On the asymptotic efficiency on moment and maximum likelihood estimators in the three-parameter inverse Gaussian distribution. *Commun. Statist. Theory. Meth.*, *13*:2307–2314.

Jorgensen, B. (1980). Statistical properties of the generalized inverse Gaussian distribution. *Lecture Notes in Statistics, Vol. 9*, New York: Springer-Verlag.

Karlin, S. (1968). *Total Positivity*, Vol. I. Stanford: Calif.: Stanford University Press.

Karlin, S. (1969). *A First Course in Stochastic Processes.* New York: Academic Press.

Kempthorne, O., and Folks, J. L. (1971). *Probability, Statistics, and Data Analysis.* Ames: Iowa State University Press.

Kendall, D. G. (1957). Some Problems in the Theory of Dams, *J. Roy. Statist. Soc.*, *B19*:207–212.

Kenney, J. F., and Keeping, E. S. (1954). *Mathematics of Statistics*, Part One, New York: Van Nostrand.

Khatri, C. G. (1962). A characterization of the inverse Gaussian distribution. *Ann. Math. Statist.*, *33*:800–803.

Kirmani, S. N. U. A., and Ahsanullah, M. (1986). A note on weighted distributions. *Commun. Statist. Theor. Meth.* (accepted for publication).

Korwar, R. M. (1980). On the uniformly minimum variance unbiased estimators of the variance and its reciprocal of an inverse Gaussian distribution. *J. Amer. Statist. Ass.*, *75*:734–735.

Lancaster, A. (1972). A stochastic model for the duration of a strike. *J. R. Statist. Soc.*, *A135*:257–271.

Lawless, J. F. (1982). *Statistical Models and Methods for Lifetime Data*, New York: Wiley.

Lawless, J. F. (1983). Statistical methods in reliability (with discussions). *Technometrics*, *25*:305–335.

Lehmann, E. L. (1959). *Testing Statistical Hypotheses.* New York: Wiley.

Letac, G., and Seshadri, V. (1983). A characterization of the generalized inverse Gaussian distribution by continued fractions. *Z. Wahrscheinlichkeitstheorie Verw. Geb.*, *62*:485–489.

Letac, G., and Seshadri, V. (1986). On a conjecture concerning inverse Gaussian regression. *Int. Statist. Rev.*, *54*:187–190.

Letac, G., Seshadri, V., and Whitmore, G. A. (1985). An exact chi-squared decomposition theorem for inverse Gaussian variates. *J. Roy. Statist. Soc.*, *B47*:476–481.

Levy, P. (1948). *Processes Stochastiques et Mouvement Brownian.* Paris: Gauthier-Villars.

Lieblein, J., and Zelen, M. (1956). Statistical investigation of the fatigue life of deep-groove ball bearings. *J. Res. Nat. Bur. Stand.*, *57:*273–316.

Lingappaiah, G. S. (1983). Prediction in samples from the inverse Gaussian distribution. *Statistica*, *43*:259–265.

Lombard, F. (1978). A sequential test for the mean of an inverse Gaussian distribution. *South African Statist. J.*, *12*:107–115.

MacDonald, D. K. C. (1962). *Noise and Fluctuations.* New York: Wiley.

Marcus, A. H. (1975). Some exact distributions in traffic noise theory. *Adv. Appl. Prob.*, *7*:593–606.

Marcus, A. H. (1975). Power sum distributions: an easier approach using the Wald distribution. *J. Amer. Statist. Ass.*, *71*:237–238.

Marcus, A. H. (1975). Power laws in compartmental analysis. Part I: a unified stochastic model. *Math. Biosci.*, *23*:337–350.

Mardia, K. V. (1970). *Families of Bivariate Distributions*, Darien, Conn.: Hafner Publishing Co.

Martz, H. F., and Waller, R. A. (1982). *Bayesian Reliability Analysis.* New York: Wiley.

Maxim, D. L., and Harrington, L. (1982). *Photogrammetric Eng. Remote Sensing*, *48*: 1271–1287.

McCool, J. I. (1979). Analysis of single classification experiments based on censored samples from the two-parameter Weibull distribution. *J. Stat. Plan. Inf.*, *3*:39–68.

Michael, J. R., Schucany, W. R., and Haas, R. W. (1976). Generating random variables using transformation with multiple roots. *Amer. Statist.*, *30*:88–90.

Miura, C. K. (1978). Tests for the mean of the inverse Gaussian distribution. *Scand. J. Statist. 5*:200–204.

Moran, P. A. P. (1968). *An Introduction to Probability Theory.* Oxford: Clarendon Press.

Mukhopadhyay, N. (1982). Stein's two-stage procedure and exact consistency. *Scand. Actuarial J.*, 110–122.

Nadas, A. (1973). Best tests for zero drift based on first passage times in Brownian motion. *Technometrics, 15*:125–132.

Nelson, W. B. (1971). *IEEE Trans. Electrical Insulations, EI-6*: 165–181.

Ord, J. K. (1975). Statistical models for personal income distributions, pp. 151–158. in *Statistical Distributions in Scientific Work, Vol. 2: Model Building and Model Selection*, edited by G. P. Patil et al. Dordrecht: Reidel.

Padgett, W. J. (1979). Confidence bounds on reliability for the inverse Gaussian model. *IEEE Trans. Reliab., R-28*:165–168.

Padgett, W. J. (1981). Bayes estimation of reliability for the inverse Gaussian model. *IEEE Trans. Reliab., R-30*:384–385.

Padgett, W. J. (1982). An approximate prediction interval for the mean of future observations from the inverse Gaussian distribution. *J. Statist. Comput. Simul. 14*:191–199.

Padgett, W. J., and Wei, L. J. (1979). Estimation for the three-parameter inverse Gaussian distribution, *Comm. Statist., A8*:129–137.

Padmanabhan, P. (1978). Applications of the inverse Gaussian distribution in evaluation and estimation of conversion probabilities for convertible securities. MBA Research Paper, McGill University, Montreal, Quebec.

Palmer, J. (1973). Certain non-classical inference procedures applied to the inverse Gaussian distribution. Ph.D. Dissertation, Oklahoma State University, Stillwater.

Patel, R. C. (1965). Estimates of parameters of truncated inverse Gaussian distribution. *Ann. Inst. Statist. Math., 17*:29–33.

Patil, S. A., and Kovner, J. L. (1976). On the test and power of zero drift on first passage times in Brownian motion. *Technometrics, 18*:341–342.

Patil, S. A., and Kovner, J. L. (1979). On the power of an optimum test for the mean of the inverse Gaussian distribution. *Technometrics, 21*:379–381.

Prabhu, N. U. (1965). *Stochastic Processes*. New York: Macmillan.

Ramachandran, G. (1975). Extreme order statistics in large samples from exponential type distributions and their application to fire loss. pp. 355–367. In *Statistical Distributions in Scientific Work, Vol. 2: Model Building and Model Selection*, edited by G. P. Patil et al. Dordrecht: Reidel.

Rodieck, R. W., Gerstein, G. L. and Kiang, N. Y. S. (1962). Some quantitative methods for the study of spontaneous activity of single neurons. *Biophys. J.*, *2*:351–367.

Roy, L. K. (1970). Estimation of the parameter of the inverse Gaussian distribution from a multi-censored sample. *Statistica, 30*:563–567.

Roy, L. K., and Wasan, M. T. (1968a). The first passage time distribution of Brownian motion with positive drift. *Math. Biosci., 3*:191–204.

Roy, L. K., and Wasan, M. T. (1968b). Properties of the time distribution of standard Brownian motion. *Trabajos Estadist., 19*:1–11.

Roy, L. K., and Wasan, M. T. (1968c). Some characteristic properties of the time distribution of standard Brownian motion. *Recu en Mars*, 29–38.

Roy, L. K., and Wasan, M. T. (1969). A characterization of the inverse Gaussian distribution. *Sankhya A31*:217–218.

Shaban, S. A. (1981). Computation of the Poisson-inverse Gaussian distribution. *Commun. Statist., B*, 1389–1399.

Sankaran, M. (1968). Mixtures by the inverse Gaussian distribution. *Sankhyā, B30*:455–458.

Schrodinger, E. (1915). Zur theorie der fall- und steigversuche an teilchen mit Brownscher bewegung. *Phys. Ze., 16*:289–295.

Seal, H. L. (1969). *Stochastic Theory of a Risk Business*. New York: Wiley.

Seal, H. L. (1978). *Astin Bull., 10*: 47–53.

Seshadri, V. (1981). A note on the inverse Gaussian distribution, in *Statistical Distributions in Scientific Work*, Vol. 4, edited by Taillie et al. Dordrecht: Reidel, pp. 99–103.

Seshadri, V. (1983). The inverse Gaussian distribution: some properties and characterizations. *Canad. J. Statist., 11*:131–136.

Seshadri, V., and Shuster, J. J. (1974). Exact tests for zero drift based on first passage times in Brownian motion. *Technometrics, 16*:133–134.

Shaban, S. A. (1981). On the discrete Poisson-inverse Gaussian distribution. *Biom. J., 23*:297–303.

Shen, C.-L. (1936). Fundamental of the theory of inverse sampling. *Ann. Math. Statist.*, *6*:62–112.

Sheppard, C. W. (1962). *Basic Principles of the Tracer Method.* New York: Wiley.

Sheppard, C. W., and Savage, L. J. (1951). The random walk problem in relation to the physiology of circulatory mixing. *Phys. Rev.*, *83*:489–490.

Shuster, J. J. (1968). On the inverse Gaussian distribution function. *J. Amer. Statist. Ass.*, *63*:1514–1516.

Shuster, J. J., and Miura, C. (1972). Two way analysis of reciprocals. *Biometrika*, *59*:478–481.

Sichel, H. S. (1971). On a family of discrete distributions particularly suited to represent long tailed frequency data. *Proceedings of the Third Symposium on Mathematical Statistics*, S.A. CSIR (Pretoria), pp. 51–97.

Sichel, H. S. (1974). On a distribution representing sentence-length in written prose. *J. Roy. Statist. Soc.*, *A137*:25–34.

Sichel, H. S. (1975). On a distribution law for word frequencies. *J. Amer. Statist. Ass.*, *70*:542–547.

Sichel, H. S. (1982). Asymptotic efficiencies of three methods of estimation for the inverse Gaussian-Poisson distribution. *Biometrika*, *67*:467–472.

Silcock, H. (1954). The phenomenon of labour turnover. *J. Roy Statist. Soc.*, *A117*:429–440.

Smoluchowski, M. V. (1915). Notiz uber die berechnung der Browschen molekular-bewegung bei der ehrenhaft-millikanschen versuchsanordnung. *Phy. Z.*, *16*:318–321.

Stark, G. E., and Chhikara, R. S. (1988). Kolmogorov-Smirnov statistics for the inverse gaussian distribution. Presented at the IASTED International Conference on Applied Simulation and Modelling, held in Galveston, Texas, May 18–20, 1988. Also to appear in the Conference Proceedings.

Stephens, M. A. (1974). EDF statistics for goodness of fit and some comparisons. *J. Amer. Statist. Assoc.*, *69*:730–737.

Stewart, D. A., and Essenwanger, O. M. (1978). Frequency distribution of wind speed near the surface. *Appl. Meteorol.*, *17*:1633–1642.

Tweedie, M. C. K. (1941). A mathematical investigation of some electrophoretic measurements on colloids, M.Sc. Thesis, University of Reading, England.

Tweedie, M. C. K. (1946). The regression of the sample variance on the sample mean. *J. Lond. Math. Soc.*, *21*:22–28.

Tweedie, M. C. K. (1945). Inverse statistical variates. *Nature, 155*:453.

Tweedie, M. C. K. (1956). Some statistical properties of inverse Gaussian distributions. *Virginia J. Sci.*, *7*:160–165.

Tweedie, M. C. K. (1957a). Statistical properties of inverse Gaussian distributions I. *Ann. Math. Statist.*, *28*:362–377.

Tweedie, M. C. K. (1957b). Statistical properties of inverse Gaussian distributions II. *Ann. Math. Statist.*, *28*:696–705.

Von Alven, W. H. (ed.), (1964). *Reliability Engineering* by ARINC, Englewood Cliffs, N.J.: Prentice-Hall, Inc.

Wald, A. (1944). On cumulative sums of random variables. *Ann. Math. Statist.*, *15*:283–296.

Wald, A. (1945). Sequential tests of statistical hypotheses. *Ann. Math. Statist.*, *16*:117–186.

Wald, A. (1947). *Sequential Analysis*, New York: Wiley.

Wasan, M. T. (1968). On an inverse Gaussian process. *Skand. Actuar.*, *60*:69–96.

Wasan, M. T. (1969a). First passage time distribution of Brownian motion with positive drift (inverse Gaussian distribution). Queen's Paper in Pure and Applied Mathematics, 19, Queens University, Canada.

Wasan, M. T. (1969b). Sufficient conditions for a first passage time process to be that of Brownian motion. *J. Appl. Probab.*, *6*:218–223.

Wasan, M. T., and Roy, L. K. (1969). Tables of inverse Gaussian percentage points. *Technometrics*, *11*:591–604.

Watson, G. N. (1966). *A Treatise on the Theory of Bessel Functions*, 2nd ed. Cambridge: Cambridge University Press.

Watson, G. S., and Wells, W. T. (1961). On the possibility of improving the mean useful life of items by eliminating those with short lives. *Technometrics, 3*:281–298.

Whitmore, G. A. (1976). Management applications of the inverse Gaussian distributions. *Int. J. Manage. Sci., 4*:215–223.

Whitmore, G. A. (1979). An inverse Gaussian model for labour turnover. *J. Roy. Statist. Soc., A, 142*:468–478.

Whitmore, G. A. (1983). A regression method for censored inverse-Gaussian data. *Can. J. Statist., 11*:305–315.

Whitmore, G. A. (1986a). Inverse Gaussian ratio estimation. *Appl. Statist., 35*:8–15.

Whitmore, G. A. (1986b). First-passage-time models for duration data: regression structures and competing risks. *The Statistician, 35*:207–219.

Whitmore, G. A., and Seshadri, Y. (1987). *Am. Statist., 41*: 280–281.

Whitmore, G. A., and Yalovsky, M. (1978). A normalizing logarithmic transformation for inverse Gaussian random variables. *Technometrics, 20*:207–208.

Wiener, N. (1923). Differential space. *J. Math. Phys., 2*:131–174.

Wise, M. E. (1966). Tracer dilution curves in cardiology and random walks and lognormal distributions. *Acta Phys. Pharmacol. Neerl., 14*:175–204.

Wise, M. E. (1971). Skew probability curves with negative powers of time and related to random walks in series, *Statist. Neerl., 25*:159–180.

Wise, M. E. (1975). Skew distributions in biomedicine including some with negative powers of time, in *Statistical Distributions in Scientific Work, Vol. 2: Model Building and Model Selection* (G. P. Patil et al., eds). Dordrecht: Reidel, pp. 241–262.

Wise, M. E., Osborn, S. B., Anderson, J., and Tomlinson, R. W. S. (1968). A stochastic model for turnover of radiocalcium based on the observed laws. *Math. Biosci., 2*:199–224.

Yule, G. Y. (1944). *The Statistical Study of Literary Vocabulary.* Cambridge: Cambridge University Press.

Zigangirov, K. S. (1962). Expression for the Wald distribution in terms of normal distribution. *Radiotekhn. Electron., 7*:164–166.

Index

Printed and bound by CPI Group (UK) Ltd, Croydon, CR0 4YY

23/10/2024

01778237-0003